函数式 Python 编程 第 2 版（影印版）
Functional Python Programming, 2nd Edition

Steven F. Lott 著

南京　东南大学出版社

图书在版编目(CIP)数据

函数式 Python 编程:第 2 版:英文/(美)史帝文·F. 洛特(Steven F. Lott)著. 一影印本. 一南京:东南大学出版社,2019.5

书名原文:Functional Python Programming, 2nd Edition

ISBN 978-7-5641-8325-7

Ⅰ. ①函… Ⅱ. ①史… Ⅲ. ①软件工具-程序设计-英文 Ⅳ. ①TP311.561

中国版本图书馆 CIP 数据核字(2019)第 046188 号

图字:10-2018-499 号

函数式 Python 编程 第 2 版(影印版)

出版发行:东南大学出版社
地　　址:南京四牌楼 2 号　　邮编:210096
出 版 人:江建中
网　　址:http://www.seupress.com
电子邮件:press@seupress.com
印　　刷:常州市武进第三印刷有限公司
开　　本:787 毫米×980 毫米　　16 开本
印　　张:25.5
字　　数:499 千字
版　　次:2019 年 5 月第 1 版
印　　次:2019 年 5 月第 1 次印刷
书　　号:ISBN 978-7-5641-8325-7
定　　价:99.00 元

本社图书若有印装质量问题,请直接与营销部联系。电话(传真):025-83791830

About the author

Steven F. Lott has been programming since the '70s, when computers were large, expensive, and rare. He's been using Python to solve business problems for over 10 years. His other titles with Packt Publishing include *Python Essentials, Mastering Object-Oriented Python, Functional Python Programming,* and *Python for Secret Agents*. Steven is currently a technomad who lives in city along the east coast of the U.S. You can follow his technology blog (*slott-softwarearchitect*).

About the reviewer

Yogendra Sharma is a developer with experience in architecture, design, and development of scalable and distributed applications. He was awarded a bachelor's degree from the Rajasthan Technical University in computer science. With a core interest in microservices and Spring, he also has hands-on experience in technologies such as AWS Cloud, Python, J2EE, NodeJS, JavaScript, Angular, MongoDB, and Docker.

Currently, he works as an IoT and Cloud Architect at Intelizign Engineering Services Pune.

Packt is searching for authors like you

If you're interested in becoming an author for Packt, please visit `authors.packtpub.com` and apply today. We have worked with thousands of developers and tech professionals, just like you, to help them share their insight with the global tech community. You can make a general application, apply for a specific hot topic that we are recruiting an author for, or submit your own idea.

Table of Contents

Preface

Functional programming offers a variety of techniques for creating succinct and expressive software. While Python is not a purely functional programming language, we can do a great deal of functional programming in Python.

Python has a core set of functional programming features. This lets us borrow many design patterns and techniques from other functional languages. These borrowed concepts can lead us to create succinct and elegant programs. Python's generator expressions, in particular, negate the need to create large in-memory data structures, leading to programs that may execute more quickly because they use fewer resources.

We can't easily create purely functional programs in Python. Python lacks a number of features that would be required for this. We don't have unlimited recursion, for example, we don't have lazy evaluation of all expressions, and we don't have an optimizing compiler.

There are several key features of functional programming languages that are available in Python. One of the most important ones is the idea of functions being first-class objects. Python also offers a number of higher-order functions. The built-in `map()`, `filter()`, and `functools.reduce()` functions are widely used in this role, and less-obvious are functions such as `sorted()`, `min()`, and `max()`.

We'll look at the core features of functional programming from a Python point of view. Our objective is to borrow good ideas from functional programming languages and use those ideas to create expressive and succinct applications in Python.

Who this book is for

This book is for programmers who want to create succinct, expressive Python programs by borrowing techniques and design patterns from functional programming languages. Some algorithms can be expressed elegantly in a functional style; we can—and should—adapt this to make Python programs more readable and maintainable.

In some cases, a functional approach to a problem will also lead to extremely high-performance algorithms. Python makes it too easy to create large intermediate data structures, tying up memory (and processor time.) With functional programming design patterns, we can often replace large lists with generator expressions that are equally expressive but take up much less memory and run much more quickly.

What this book covers

`Chapter 1`, *Understanding Functional Programming*, introduces some of the techniques that characterize functional programming. We'll identify some of the ways to map those features to Python. Finally, we'll also address some ways that the benefits of functional programming accrue when we use these design patterns to build Python applications.

`Chapter 2`, *Introducing Essential Functional Concepts*, delves into six central features of the functional programming paradigm. We'll look at each in some detail to see how they're implemented in Python. We'll also point out some features of functional languages that don't apply well to Python. In particular, many functional languages have complex type-matching rules required to support compiling and optimizing.

`Chapter 3`, *Functions, Iterators, and Generators*, will show how to leverage immutable Python objects, and generator expressions adapt functional programming concepts to the Python language. We'll look at some of the built-in Python collections and how we can leverage them without departing too far from functional programming concepts.

`Chapter 4`, *Working with Collections*, shows how you can use a number of built-in Python functions to operate on collections of data. This chapter will focus on a number of relatively simple functions, such as `any()` and `all()`, which will reduce a collection of values to a single result.

`Chapter 5`, *Higher-Order Functions*, examines the commonly-used higher-order functions such as `map()` and `filter()`. It also shows a number of other functions that are also higher-order functions as well as how we can create our own higher-order functions.

`Chapter 6`, *Recursions and Reductions*, teaches how to design an algorithm using recursion and then optimize it into a high-performance `for` loop. We'll also look at some other reductions that are widely used, including `collections.Counter()`.

`Chapter 7`, *Additional Tuple Techniques*, showcases a number of ways that we can use immutable tuples (and namedtuples) instead of stateful objects. Immutable objects have a much simpler interface—we never have to worry about abusing an attribute and setting an object into some inconsistent or invalid state.

`Chapter 8`, *The Itertools Module*, examines a number of functions in this standard library module. This collection of functions simplifies writing programs that deal with collections or generator functions.

`Chapter 9`, *More Itertools Techniques*, covers the combinatoric functions in the itertools module. These functions are somewhat less useful. This chapter includes some examples that illustrate ill-considered use of these functions and the consequences of combinatoric explosion.

`Chapter 10`, *The Functools Module*, focuses on how to use some of the functions in this module for functional programming. A few functions in this module are more appropriate for building decorators, and they are left for `Chapter 11`, *Decorator Design Techniques*. The other functions, however, provide several more ways to design and implement function programs.

`Chapter 11`, *Decorator Design Techniques*, looks at how you can look at a decorator as a way to build a composite function. While there is considerable flexibility here, there are also some conceptual limitations: we'll look at ways that overly-complex decorators can become confusing rather than helpful.

`Chapter 12`, *The Multiprocessing and Threading Modules*, points out an important consequence of good functional design: we can distribute the processing workload. Using immutable objects means that we can't corrupt an object because of poorly-synchronized write operations.

`Chapter 13`, *Conditional Expressions and the Operator Module*, lists some ways to break out of Python's strict order of evaluation. There are limitations to what we can achieve here. We'll also look at the operator module and how this can lead to slight clarification of some simple kinds of processing.

`Chapter 14`, *The PyMonad Library*, examines some of the features of the PyMonad library. This provides some additional functional programming features. It also provides a way to learn more about monads. In some functional languages, monads are an important way to force a particular order for operations that might get optimized into an undesirable order. Since Python already has strict ordering of *f* expressions and statements, the monad feature is more instructive than practical.

`Chapter 15`, *A Functional Approach to Web Services*, shows how we can think of web services as a nested collection of functions that transform a request into a reply. We'll see ways to leverage functional programming concepts for building responsive, dynamic web content.

`Chapter 16`, *Optimizations and Improvements*, includes some additional tips on performance and optimization. We'll emphasize techniques such as memoization, because they're easy to implement and can—in the right context—yield dramatic performance improvements.

To get the most out of this book

This book presumes some familiarity with Python 3 and general concepts of application development. We won't look deeply at subtle or complex features of Python; we'll avoid much consideration of the internals of the language.

We'll presume some familiarity with functional programming. Since Python is not a functional programming language, we can't dig deeply into functional concepts. We'll pick and choose the aspects of functional programming that fit well with Python and leverage just those that seem useful.

Some of the examples use **exploratory data analysis** (**EDA**) as a problem domain to show the value of functional programming. Some familiarity with basic probability and statistics will help with this. There are only a few examples that move into more serious data science.

You'll need to have Python 3.6 installed and running. For more information on Python, visit `http://www.python.org/`. The examples all make extensive use of type hints, which means that the latest version of **mypy** must be installed as well.

Check out `https://pypi.python.org/pypi/mypy` for the latest version of **mypy**.

Examples in `Chapter 9`, *More Itertools Techniques,* use PIL and Beautiful Soup 4. The Pillow fork of the original PIL library works nicely; refer to `https://pypi.python.org/pypi/Pillow/2.7.0` and `https://pypi.python.org/pypi/beautifulsoup4/4.6.0`.

Examples in `Chapter 14`, *The PyMonad Library,* use PyMonad; check out `https://pypi.python.org/pypi/PyMonad/1.3`.

All of these packages should be installed using the following:

```
$ pip install pillow beautifulsoup4 PyMonad
```

Download the example code files

You can download the example code files for this book from your account at `www.packtpub.com`. If you purchased this book elsewhere, you can visit `www.packtpub.com/support` and register to have the files emailed directly to you.

You can download the code files by following these steps:

1. Log in or register at `www.packtpub.com`.
2. Select the **SUPPORT** tab.
3. Click on **Code Downloads & Errata**.
4. Enter the name of the book in the **Search** box and follow the onscreen instructions.

Once the file is downloaded, please make sure that you unzip or extract the folder using the latest version of:

- WinRAR/7-Zip for Windows
- Zipeg/iZip/UnRarX for Mac
- 7-Zip/PeaZip for Linux

The code bundle for the book is also hosted on GitHub at `https://github.com/PacktPublishing/Functional-Python-Programming-Second-Edition/`. We also have other code bundles from our rich catalog of books and videos available at `https://github.com/PacktPublishing/`. Check them out!

Conventions used

There are a number of text conventions used throughout this book.

`CodeInText`: Indicates code words in text, database table names, folder names, filenames, file extensions, pathnames, dummy URLs, user input, and Twitter handles. Here is an example: "Python has other statements, such as `global` or `nonlocal`, which modify the rules for variables in a particular namespace."

A block of code is set as follows:

```
s = 0
for n in range(1, 10):
    if n % 3 == 0 or n % 5 == 0:
        s += n
print(s)
```

When we wish to draw your attention to a particular part of a code block, the relevant lines or items are set in bold:

```
s = 0
for n in range(1, 10):
    if n % 3 == 0 or n % 5 == 0:
        s += n
print(s)
```

Any command-line input or output is written as follows:

```
$ pip install pillow beautifulsoup4 PyMonad
```

Bold: Indicates a new term, an important word, or words that you see onscreen. For example, words in menus or dialog boxes appear in the text like this. Here is an example: "For our purposes, we will distinguish between only two of the many paradigms: **functional programming** and **imperative programming**."

Warnings or important notes appear like this.

Tips and tricks appear like this.

Get in touch

Feedback from our readers is always welcome.

General feedback: Email `feedback@packtpub.com` and mention the book title in the subject of your message. If you have questions about any aspect of this book, please email us at `questions@packtpub.com`.

Errata: Although we have taken every care to ensure the accuracy of our content, mistakes do happen. If you have found a mistake in this book, we would be grateful if you would report this to us. Please visit `www.packtpub.com/submit-errata`, selecting your book, clicking on the Errata Submission Form link, and entering the details.

Piracy: If you come across any illegal copies of our works in any form on the Internet, we would be grateful if you would provide us with the location address or website name. Please contact us at `copyright@packtpub.com` with a link to the material.

If you are interested in becoming an author: If there is a topic that you have expertise in and you are interested in either writing or contributing to a book, please visit `authors.packtpub.com`.

Reviews

Please leave a review. Once you have read and used this book, why not leave a review on the site that you purchased it from? Potential readers can then see and use your unbiased opinion to make purchase decisions, we at Packt can understand what you think about our products, and our authors can see your feedback on their book. Thank you!

For more information about Packt, please visit `packtpub.com`.

1
Understanding Functional Programming

Functional programming defines a computation using expressions and evaluation; often these are encapsulated in function definitions. It de-emphasizes or avoids the complexity of state change and mutable objects. This tends to create programs that are more succinct and expressive. In this chapter, we'll introduce some of the techniques that characterize functional programming. We'll identify some of the ways to map these features to **Python**. Finally, we'll also address some ways in which the benefits of functional programming accrue when we use these design patterns to build Python applications.

Python has numerous functional programming features. It is not a purely a functional programming language. It offers enough of the right kinds of features that it confers the benefits of functional programming. It also retains all the optimization power of an imperative programming language.

We'll also look at a problem domain that we'll use for many of the examples in this book. We'll try to stick closely to **Exploratory Data Analysis** (**EDA**) because its algorithms are often good examples of functional programming. Furthermore, the benefits of functional programming accrue rapidly in this problem domain.

Our goal is to establish some essential principles of functional programming. The more serious Python code will begin in `Chapter 2`, *Introducing Some Functional Features*.

We'll focus on Python 3.6 features in this book. However, some of the examples might also work in Python 2.

Identifying a paradigm

It's difficult to be definitive on the universe of programming paradigms. For our purposes, we will distinguish between only two of the many paradigms: **functional programming** and **imperative programming**. One important distinguishing feature between these two is the concept of **state**.

In an imperative language, such as Python, the state of the computation is reflected by the values of the variables in the various namespaces; some kinds of statements make a well-defined change to the state by adding or changing (or even removing) a variable. A language is imperative because each statement is a command, which changes the state in some way.

Our general focus is on the assignment statement and how it changes the state. Python has other statements, such as `global` or `nonlocal`, which modify the rules for variables in a particular namespace. Statements such as `def`, `class`, and `import` change the processing context. Other statements such as `try`, `except`, `if`, `elif`, and `else` act as guards to modify how a collection of statements will change the computation's state. Statements such as `for` and `while`, similarly, wrap a block of statements so that the statements can make repeated changes to the state of the computation. The focus of all these various statement types, however, is on changing the state of the variables.

Ideally, each assignment statement advances the state of the computation from an initial condition toward the desired final outcome. This *advancing the computation* assertion can be challenging to prove. One approach is to define the final state, identify a statement that will establish this final state, and then deduce the precondition required for this final statement to work. This design process can be iterated until an acceptable initial state is derived.

In a functional language, we replace the state—the changing values of variables—with a simpler notion of evaluating functions. Each function evaluation creates a new object or objects from existing objects. Since a functional program is a composition of functions, we can design lower-level functions that are easy to understand, and then design higher-level compositions that can also be easier to visualize than a complex sequence of statements.

Function evaluation more closely parallels mathematical formalisms. Because of this, we can often use simple algebra to design an algorithm, which clearly handles the edge cases and boundary conditions. This makes us more confident that the functions work. It also makes it easy to locate test cases for formal unit testing.

It's important to note that functional programs tend to be relatively succinct, expressive, and efficient compared to imperative (object-oriented or procedural) programs. The benefit isn't automatic; it requires a careful design. This design effort for functional programming is often easier than for procedural programming.

Subdividing the procedural paradigm

We can subdivide imperative languages into a number of discrete categories. In this section, we'll glance quickly at the procedural versus object-oriented distinction. What's important here is to see how object-oriented programming is a subset of imperative programming. The distinction between procedural and object-orientation doesn't reflect the kind of fundamental difference that functional programming represents.

We'll use code examples to illustrate the concepts. For some, this will feel like reinventing the wheel. For others, it provides a concrete expression of abstract concepts.

For some kinds of computations, we can ignore Python's object-oriented features and write simple numeric algorithms. For example, we might write something like the following to sum a range of numbers that share a common property:

```
s = 0
for n in range(1, 10):
    if n % 3 == 0 or n % 5 == 0:
        s += n
print(s)
```

The sum `s` includes only numbers that are multiples of three or five. We've made this program strictly procedural, avoiding any explicit use of Python's object features. The program's state is defined by the values of the variables `s` and `n`. The variable `n` takes on values such that $1 \le n < 10$. As the loop involves an ordered exploration of values of `n`, we can prove that it will terminate when `n == 10`. Similar code would work in C or Java language, using their primitive (non-object) data types.

We can exploit **Python's Object-Oriented Programming** (**OOP**) features and create a similar program:

```
m = list()
for n in range(1, 10):
    if n % 3 == 0 or n % 5 == 0:
        m.append(n)
print(sum(m))
```

This program produces the same result but it accumulates a stateful collection object, `m`, as it proceeds. The state of the computation is defined by the values of the variables `m` and `n`.

The syntax of `m.append(n)` and `sum(m)` can be confusing. It causes some programmers to insist (wrongly) that Python is somehow not purely object-oriented because it has a mixture of the `function()` and `object.method()` syntax. Rest assured, Python is purely object-oriented. Some languages, such as C++, allow the use of primitive data types such as `int`, `float`, and `long`, which are not objects. Python doesn't have these primitive types. The presence of prefix syntax, `sum(m)`, doesn't change the nature of the language.

To be pedantic, we could fully embrace the object model, by defining a subclass of the `list` class. This new class will include a `sum` method:

```
class Summable_List(list):
    def sum(self):
        s = 0
        for v in self:
            s += v
        return s
```

If we initialize the variable `m` with an instance of the `Summable_List()` class instead of the `list()` method, we can use the `m.sum()` method instead of the `sum(m)` method. This kind of change can help to clarify the idea that Python is truly and completely object-oriented. The use of prefix function notation is purely syntactic sugar.

All three of these examples rely on variables to explicitly show the state of the program. They rely on the assignment statements to change the values of the variables and advance the computation toward completion. We can insert the `assert` statements throughout these examples to demonstrate that the expected state changes are implemented properly.

The point is not that imperative programming is broken in some way. The point is that functional programming leads to a change in viewpoint, which can, in many cases, be very helpful. We'll show a function view of the same algorithm. Functional programming doesn't make this example dramatically shorter or faster.

Using the functional paradigm

In a functional sense, the sum of the multiples of three and five can be defined in two parts:

- The sum of a sequence of numbers
- A sequence of values that pass a simple test condition, for example, being multiples of three and five

The sum of a sequence has a simple, recursive definition:

```
def sumr(seq):
    if len(seq) == 0: return 0
    return seq[0] + sumr(seq[1:])
```

We've defined the sum of a sequence in two cases: the **base case** states that the sum of a zero length sequence is 0, while the **recursive case** states that the sum of a sequence is the first value plus the sum of the rest of the sequence. Since the recursive definition depends on a shorter sequence, we can be sure that it will (eventually) devolve to the base case.

Here are some examples of how this function works:

```
>>> sumr([7, 11])
18
>>> 7+sumr([11])
18
>>> 18+sumr([])
0
```

The first example computes the sum of a list with multiple items. The second example shows how the recursion rule works by adding the first item, `seq[0]`, to the sum of the remaining items, `sumr(seq[1:])`. Eventually, the computation of the result involves the sum of an empty list, which is defined as zero.

The `+` operator on the last line of the preceding example and the initial value of `0` in the base case characterize the equation as a sum. If we change the operator to `*` and the initial value to `1`, it would just as easily compute a product. We'll return to this simple idea of generalization in the following chapters.

Similarly, a sequence of values can have a simple, recursive definition, as follows:

```
def until(n, filter_func, v):
    if v == n: return []
    if filter_func(v): return [v] + until(n, filter_func, v+1)
    else: return until(n, filter_func, v+1)
```

In this function, we've compared a given value, `v`, against the upper bound, `n`. If `v` reaches the upper bound, the resulting list must be empty. This is the base case for the given recursion.

There are two more cases defined by the given `filter_func()` function. If the value of `v` is passed by the `filter_func()` function, we'll create a very small list, containing one element, and append the remaining values of the `until()` function to this list. If the value of `v` is rejected by the `filter_func()` function, this value is ignored and the result is simply defined by the remaining values of the `until()` function.

We can see that the value of `v` will increase from an initial value until it reaches `n`, assuring us that we'll reach the base case soon.

Here's how we can use the `until()` function to generate the multiples of three and five. First, we'll define a handy `lambda` object to filter values:

```
mult_3_5 = lambda x: x%3==0 or x%5==0
```

(We will use lambdas to emphasize succinct definitions of simple functions. Anything more complex than a one-line expression requires the `def` statement.)

We can see how this lambda works from Command Prompt in the following example:

```
>>> mult_3_5(3)
True
>>> mult_3_5(4)
False
>>> mult_3_5(5)
True
```

This function can be used with the `until()` function to generate a sequence of values, which are multiples of three and five.

The `until()` function for generating a sequence of values works as follows:

```
>>> until(10, lambda x: x%3==0 or x%5==0, 0)
[0, 3, 5, 6, 9]
```

We can use our recursive `sum()` function to compute the sum of this sequence of values. The various functions such as `sum()`, `until()`, and `mult_3_5()` are defined as simple recursive functions. The values are computed without resorting to using intermediate variables to store the state.

We'll return to the ideas behind this purely functional, recursive definition in several places. It's important to note here that many functional programming language compilers can optimize these kinds of simple recursive functions. Python can't do the same optimizations.

Using a functional hybrid

We'll continue this example with a mostly functional version of the previous example to compute the sum of multiples of three and five. Our **hybrid** functional version might look like the following:

```
print(sum(n for n in range(1, 10) if n%3==0 or n%5==0))
```

We've used nested generator expressions to iterate through a collection of values and compute the sum of these values. The `range(1, 10)` method is iterable and, consequently, a kind of generator expression; it generates a sequence of values $\{n|1 \leq n < 10\}$. The more complex expression `n for n in range(1, 10) if n%3==0 or n%5==0` is also an iterable expression. It produces a set of values, $\{n|1 \leq n < 10 \wedge (n \mod 3 = 0 \vee n \mod 5 = 0\}$. The variable `n` is bound to each value, more as a way of expressing the contents of the set than as an indicator of the state of the computation. The `sum()` function consumes the iterable expression, creating a final object, 23.

The bound variable doesn't exist outside the generator expression. The variable `n` isn't visible elsewhere in the program.

The `if` clause of the expression can be extracted into a separate function, allowing us to easily repurpose this for other rules. We could also use a higher-order function named `filter()` instead of the `if` clause of the generator expression. We'll save this for `Chapter 5`, *Higher-Order Functions*.

The variable `n` in this example isn't directly comparable to the variable `n` in the first two imperative examples. A `for` statement (outside a generator expression) creates a proper variable in the local namespace. The generator expression does not create a variable in the same way as a `for` statement does:

```
>>> sum(n for n in range(1, 10) if n%3==0 or n%5==0)
23
>>> n
Traceback (most recent call last):
  File "<stdin>", line 1, in <module>
NameError: name 'n' is not defined
```

The variable `n` doesn't exist outside the binding in the generator expression. It doesn't define the state of the computation.

Looking at object creation

In some cases, it might help to look at intermediate objects as a history of the computation. What's important is that the history of a computation is not fixed. When functions are commutative or associative, then changes to the order of evaluation might lead to different objects being created. This might have performance improvements with no changes to the correctness of the results.

Consider this expression:

```
>>> 1+2+3+4
10
```

We are looking at a variety of potential computation histories with the same result. Because the + operator is commutative and associative, there are a large number of candidate histories that lead to the same result.

Of the candidate sequences, there are two important alternatives, which are as follows:

```
>>> ((1+2)+3)+4
10
>>> 1+(2+(3+4))
10
```

In the first case, we fold in values working from left to right. This is the way Python works implicitly. Intermediate objects 3 and 6 are created as part of this evaluation.

In the second case, we fold from right to left. In this case, intermediate objects 7 and 9 are created. In the case of simple integer arithmetic, the two results have identical performance; there's no optimization benefit.

When we work with something like the `list` append, we might see some optimization improvements when we change the association rules.

Here's a simple example:

```
>>> import timeit
>>> timeit.timeit("((([]+[1])+[2])+[3])+[4]")
0.8846941249794327
>>> timeit.timeit("[]+([1]+([2]+([3]+[4])))")
1.0207440659869462
```

In this case, there's some benefit to working from left to right.

What's important for functional design is the idea that the + operator (or `add()` function) can be used in any order to produce the same results. The + operator has no hidden side effects that restrict the way this operator can be used.

The stack of turtles

When we use Python for functional programming, we embark down a path that will involve a hybrid that's not strictly functional. Python is not **Haskell**, **OCaml**, or **Erlang**. For that matter, our underlying processor hardware is not functional; it's not even strictly object-oriented, CPUs are generally procedural.

> *All programming languages rest on abstractions, libraries, frameworks and virtual machines. These abstractions, in turn, may rely on other abstractions, libraries, frameworks and virtual machines. The most apt metaphor is this: the world is carried on the back of a giant turtle. The turtle stands on the back of another giant turtle. And that turtle, in turn, is standing on the back of yet another turtle.*
>
> *It's turtles all the way down.*
>
> *- Anonymous*

There's no practical end to the layers of abstractions.

More importantly, the presence of abstractions and virtual machines doesn't materially change our approach to designing software to exploit the functional programming features of Python.

Even within the functional programming community, there are both purer and less pure functional programming languages. Some languages make extensive use of `monads` to handle stateful things such as file system input and output. Other languages rely on a hybridized environment that's similar to the way we use Python. In Python, software can be generally functional, with carefully chosen procedural exceptions.

Our functional Python programs will rely on the following three stacks of abstractions:

- Our applications will be functions—all the way down—until we hit the objects
- The underlying Python runtime environment that supports our functional programming is objects—all the way down—until we hit the libraries
- The libraries that support Python are a turtle on which Python stands

The operating system and hardware form their own stack of turtles. These details aren't relevant to the problems we're going to solve.

A classic example of functional programming

As part of our introduction, we'll look at a classic example of functional programming. This is based on the paper *Why Functional Programming Matters* by John Hughes. The article appeared in a paper called *Research Topics in Functional Programming*, edited by D. Turner, published by Addison-Wesley in 1990.

Here's a link to the paper *Research Topics in Functional Programming*:

`http://www.cs.kent.ac.uk/people/staff/dat/miranda/whyfp90.pdf`

This discussion of functional programming in general is profound. There are several examples given in the paper. We'll look at just one: the Newton-Raphson algorithm for locating the roots of a function. In this case, the function is the square root.

It's important because many versions of this algorithm rely on the explicit state managed via `loops`. Indeed, the Hughes paper provides a snippet of the **Fortran** code that emphasizes stateful, imperative processing.

The backbone of this approximation is the calculation of the next approximation from the current approximation. The `next_()` function takes `x`, an approximation to the `sqrt(n)` method and calculates a next value that brackets the proper root. Take a look at the following example:

```
def next_(n, x):
    return (x+n/x)/2
```

This function computes a series of values $a_{i+1} = \frac{(a_i + n/a_i)}{2}$. The distance between the values is halved each time, so they'll quickly converge on the value such that $a = \frac{n}{a}$, which means $a = \sqrt{n}$. Note that the name `next()` would collide with a built-in function. Calling it `next_()` lets us follow the original presentation as closely as possible, using Pythonic names.

Here's how the function looks when used in Command Prompt:

```
>>> n = 2
>>> f = lambda x: next_(n, x)
>>> a0 = 1.0
>>> [round(x,4) for x in (a0, f(a0), f(f(a0)), f(f(f(a0))),)]
[1.0, 1.5, 1.4167, 1.4142]
```

We've defined the `f()` method as a `lambda` that will converge on $\sqrt{2}$. We started with 1.0 as the initial value for a_0. Then we evaluated a sequence of recursive evaluations: $a_1 = f(a_0)$, $a_2 = f(f(a_0))$, and so on. We evaluated these functions using a generator expression so that we could round off each value. This makes the output easier to read and easier to use with `doctest`. The sequence appears to converge rapidly on $\sqrt{2}$.

We can write a function, which will (in principle) generate an infinite sequence of a_i values converging on the proper square root:

```
def repeat(f, a):
    yield a
    for v in repeat(f, f(a)):
        yield v
```

This function will generate approximations using a function, `f()`, and an initial value, `a`. If we provide the `next_()` function defined earlier, we'll get a sequence of approximations to the square root of the `n` argument.

The `repeat()` function expects the `f()` function to have a single argument; however, our `next_()` function has two arguments. We can use a `lambda` object, `lambda x: next_(n, x)`, to create a partial version of the `next_()` function with one of two variables bound.

The Python generator functions can't be trivially recursive; they must explicitly iterate over the recursive results, yielding them individually. Attempting to use a simple `return repeat(f, f(a))` will end the iteration, returning a generator expression instead of yielding the sequence of values.

There are two ways to return all the values instead of returning a generator expression, which are as follows:

- We can write an explicit `for` loop as follows:

```
for x in some_iter: yield x.
```

- We can use the `yieldfrom` statement as follows:

```
yield from some_iter.
```

Both techniques of yielding the values of a recursive generator function are equivalent. We'll try to emphasize `yield from`. In some cases, however, the `yield` with a complex expression will be clearer than the equivalent mapping or generator expression.

Of course, we don't want the entire infinite sequence. It's essential to stop generating values when two values are so close to each other that either one is useful as the square root we're looking for. The common symbol for the value, which is close enough, is the Greek letter **Epsilon**, ε, which can be thought of as the largest error we will tolerate.

In Python, we have to be a little clever when taking items from an infinite sequence one at a time. It works out well to use a simple interface function that wraps a slightly more complex recursion. Take a look at the following code snippet:

```
def within(ε, iterable):
    def head_tail(ε, a, iterable):
        b = next(iterable)
        if abs(a-b) <= ε: return b
            return head_tail(ε, b, iterable)
    return head_tail(ε, next(iterable), iterable)
```

We've defined an internal function, `head_tail()`, which accepts the tolerance, ε, an item from the iterable sequence, `a`, and the rest of the iterable sequence, `iterable`. The next item from the `iterable` is bound to a name `b`. If $|a - b| \leq \epsilon$, the two values are close enough together to find the square root. Otherwise, we use the `b` value in a recursive invocation of the `head_tail()` function to examine the next pair of values.

Our `within()` function merely seeks to properly initialize the internal `head_tail()` function with the first value from the `iterable` parameter.

Some functional programming languages offer a technique that will put a value back into an `iterable` sequence. In Python, this might be a kind of `unget()` or `previous()` method that pushes a value back into the iterator. Python iterables don't offer this kind of rich functionality.

We can use the three functions `next_()`, `repeat()`, and `within()` to create a square root function, as follows:

```
def sqrt(a0, ε, n):
    return within(ε, repeat(lambda x: next_(n,x), a0))
```

We've used the `repeat()` function to generate a (potentially) infinite sequence of values based on the `next_(n,x)` function. Our `within()` function will stop generating values in the sequence when it locates two values with a difference less than ε.

When we use this version of the `sqrt()` method, we need to provide an initial seed value, `a0`, and an ε value. An expression such as `sqrt(1.0, .0001, 3)` will start with an approximation of 1.0 and compute the value of $\sqrt{3}$ to within 0.0001. For most applications, the initial `a0` value can be 1.0. However, the closer it is to the actual square root, the more rapidly this method converges.

The original example of this approximation algorithm was shown in the Miranda language. It's easy to see that there are few profound differences between Miranda and Python. The biggest difference is Miranda's ability to construct `cons`, turning a value back into an `iterable`, doing a kind of `unget`. This parallelism between Miranda and Python gives us confidence that many kinds of functional programming can be easily done in Python.

Exploratory data analysis

Later in this book, we'll use the field of **exploratory data analysis** (**EDA**) as a source for concrete examples of functional programming. This field is rich with algorithms and approaches to working with complex datasets; functional programming is often a very good fit between the problem domain and automated solutions.

While details vary from author to author, there are several widely accepted stages of EDA. These include the following:

- **Data preparation**: This might involve extraction and transformation for source applications. It might involve parsing a source data format and doing some kind of data scrubbing to remove unusable or invalid data. This is an excellent application of functional design techniques.

- **Data exploration**: This is a description of the available data. This usually involves the essential statistical functions. This is another excellent place to explore functional programming. We can describe our focus as univariate and bivariate statistics but that sounds too daunting and complex. What this really means is that we'll focus on mean, median, mode, and other related descriptive statistics. Data exploration may also involve data visualization. We'll skirt this issue because it doesn't involve very much functional programming. I'll suggest that you use a toolkit such as SciPy. Visit the following links to get more information how SciPY works and its usage:
    - `https://www.packtpub.com/big-data-and-business-intelligence/learning-scipy-numerical-and-scientific-computing`
    - `https://www.packtpub.com/big-data-and-business-intelligence/learning-python-data-visualization`
- **Data modeling and machine learning**: This tends to be proscriptive as it involves extending a model to new data. We're going to skirt around this because some of the models can become mathematically complex. If we spend too much time on these topics, we won't be able to focus on functional programming.
- **Evaluation and comparison**: When there are alternative models, each must be evaluated to determine which is a better fit for the available data. This can involve ordinary descriptive statistics of model outputs. This can benefit from functional design techniques.

The goal of EDA is often to create a model that can be deployed as a decision support application. In many cases, a model might be a simple function. A simple functional programming approach can apply the model to new data and display results for human consumption.

Summary

We've looked at programming paradigms with an eye toward distinguishing the functional paradigm from two common imperative paradigms. Our objective in this book is to explore the functional programming features of Python. We've noted that some parts of Python don't allow purely functional programming; we'll be using some hybrid techniques that meld the good features of succinct, expressive functional programming with some high-performance optimizations in Python.

In the next chapter, we'll look at five specific functional programming techniques in detail. These techniques will form the essential foundation for our hybridized functional programming in Python.

2
Introducing Essential Functional Concepts

Most of the features of functional programming are already first-class parts of Python. Our goal in writing functional Python is to shift our focus away from imperative (procedural or object-oriented) techniques to as much of an extent as possible.

We'll look at each of the following functional programming topics:

- First-class and higher-order functions, which are sometimes known as pure functions.
- Immutable data.
- Strict and non-strict evaluation. We can also call this eager versus lazy evaluation.
- Recursion instead of an explicit loop state.
- Functional type systems.

This should reiterate some concepts from the first chapter: firstly, that purely functional programming avoids the complexities of an explicit state maintained through variable assignments; secondly, that Python is not a purely functional language.

This book doesn't attempt to offer a rigorous definition of what functional programming is: Python is not a purely functional language, and a strict definition isn't helpful. Instead, we'll identify some common features that are indisputably important in functional programming. We'll steer clear of the blurry edges, and focus on the obviously functional features.

In this chapter, we'll include some of the Python 3 type hints in the examples. Type hints can help you visualize the essential purpose behind a function definition. Type hints are analyzed with the **mypy** tool. As with unit testing and static analysis with **pylint**, **mypy** can be part of a toolchain used to produce high-quality software.

First-class functions

Functional programming is often succinct and expressive. One way to achieve it is by providing functions as arguments and return values for other functions. We'll look at numerous examples of manipulating functions.

For this to work, functions must be first-class objects in the runtime environment. In programming languages such as C, a function is not a runtime object. In Python, however, functions are objects that are created (usually) by the `def` statements and can be manipulated by other Python functions. We can also create a function as a callable object or by assigning `lambda` to a variable.

Here's how a function definition creates an object with attributes:

```
>>> def example(a, b, **kw):
...     return a*b
...
>>> type(example)
<class 'function'>
>>> example.__code__.co_varnames
('a', 'b', 'kw')
>>> example.__code__.co_argcount
2
```

We've created an object, `example`, that is of the `function()` class. This object has numerous attributes. The `__code__` object associated with the function object has attributes of its own. The implementation details aren't important. What is important is that functions are first-class objects and can be manipulated just like all other objects. We previously displayed the values of two of the many attributes of a function object.

Pure functions

To be expressive, a function should be free from the confusion created by side effects. Using pure functions can also allow some optimizations by changing evaluation order. The big win, however, stems from pure functions being conceptually simpler and much easier to test.

To write a pure function in Python, we have to write local-only code. This means we have to avoid the `global` statements. We need to look closely at any use of `nonlocal`; while it is a side effect in some ways, it's confined to a nested function definition. This is an easy standard to meet. Pure functions are a common feature of Python programs.

There isn't a trivial way to guarantee a Python function is free from side effects. It is easy to carelessly break the pure function rule. If we ever want to worry about our ability to follow this rule, we could write a function that uses the `dis` module to scan a given function's `__code__.co_code` compiled code for global references. It could report on use of internal closures, and the `__code__.co_freevarstuple` method as well. This is a rather complex solution to a rare problem; we won't pursue it further.

A Python `lambda` is a pure function. While this isn't a highly recommended style, it's always possible to create pure functions through the `lambda` objects.

Here's a function created by assigning `lambda` to a variable:

```
>>> mersenne = lambda x: 2**x-1
>>> mersenne(17)
131071
```

We created a pure function using `lambda` and assigned this to the variable `mersenne`. This is a callable object with a single parameter, *x*, that returns a single value. Because lambdas can't have assignment statements, they're always pure functions and suitable for functional programming.

Higher-order functions

We can achieve expressive, succinct programs using higher-order functions. These are functions that accept a function as an argument or return a function as a value. We can use higher-order functions as a way to create composite functions from simpler functions.

Consider the Python `max()` function. We can provide a function as an argument and modify how the `max()` function behaves.

Here's some data we might want to process:

```
>>> year_cheese = [(2000, 29.87), (2001, 30.12), (2002, 30.6), (2003,
30.66),(2004, 31.33), (2005, 32.62), (2006, 32.73), (2007, 33.5),
(2008, 32.84), (2009, 33.02), (2010, 32.92)]
```

We can apply the `max()` function, as follows:

```
>>> max(year_cheese)
(2010, 32.92)
```

The default behavior is to simply compare each `tuple` in the sequence. This will return the `tuple` with the largest value on position 0.

Since the `max()` function is a higher-order function, we can provide another function as an argument. In this case, we'll use `lambda` as the function; this is used by the `max()` function, as follows:

```
>>> max(year_cheese, key=lambda yc: yc[1])
(2007, 33.5)
```

In this example, the `max()` function applies the supplied `lambda` and returns the tuple with the largest value in position 1.

Python provides a rich collection of higher-order functions. We'll see examples of each of Python's higher-order functions in later chapters, primarily in `Chapter 5`, *Higher-Order Functions.* We'll also see how we can easily write our own higher-order functions.

Immutable data

Since we're not using variables to track the state of a computation, our focus needs to stay on immutable objects. We can make extensive use of `tuples` and `namedtuples` to provide more complex data structures that are immutable.

The idea of immutable objects is not foreign to Python. There can be a performance advantage to using immutable `tuples` instead of more complex mutable objects. In some cases, the benefits come from rethinking the algorithm to avoid the costs of object mutation.

We will avoid class definitions almost entirely. It can seem like anathema to avoid objects in an **Object-Oriented Programming (OOP)** language. Functional programming simply doesn't need stateful objects. We'll see this throughout this book. There are reasons for defining `callable` objects; it is a tidy way to provide namespaces for closely related functions, and it supports a pleasant level of configurability. Also, it's easy to create a cache with a callable object, leading to important performance optimizations.

As an example, here's a common design pattern that works well with immutable objects: the `wrapper()` function. A list of tuples is a fairly common data structure. We will often process this list of tuples in one of the two following ways:

- **Using higher-order functions**: As shown earlier, we provided `lambda` as an argument to the `max()` function: `max(year_cheese, key=lambda yc: yc[1])`
- **Using the wrap-process-unwrap pattern**: In a functional context, we should call this and the `unwrap(process(wrap(structure)))` pattern

For example, look at the following command snippet:

```
>>> max(map(lambda yc: (yc[1], yc), year_cheese))[1]
(2007, 33.5)
```

This fits the three-part pattern of wrapping a data structure, finding the maximum of the wrapped structures, and then unwrapping the structure.

`map(lambda yc: (yc[1], yc), year_cheese)` will transform each item into a two-tuple with a key followed by the original item. In this example, the comparison key is merely `yc[1]`.

The processing is done using the `max()` function. Since each piece of data has been simplified to a two-tuple with position zero used for comparison, the higher-order function features of the `max()` function aren't required. The default behavior of the `max()` function uses the first item in each two-tuple to locate the largest value.

Finally, we unwrap using the subscript `[1]`. This will pick the second element of the two-tuple selected by the `max()` function.

This kind of `wrap` and `unwrap` is so common that some languages have special functions with names like `fst()` and `snd()` that we can use as function prefixes instead of a syntactic suffix of `[0]` or `[1]`. We can use this idea to modify our wrap-process-unwrap example, as follows:

```
>>> snd = lambda x: x[1]
>>> snd(max(map(lambda yc: (yc[1], yc), year_cheese)))
(2007, 33.5)
```

Here, a lambda is used to define the `snd()` function to pick the second item from a `tuple`. This provides an easier-to-read version of `unwrap(process(wrap()))`. As with the previous example, the `map(lambda... , year_cheese)` expression is used to `wrap` our raw data items, and the `max()` function does the processing. Finally, the `snd()` function extracts the second item from the tuple.

In `Chapter 13`, *Conditional Expressions and the Operator Module*, we'll look at some alternatives to `lambda` functions, such as `fst()` and `snd()`.

Strict and non-strict evaluation

Functional programming's efficiency stems, in part, from being able to defer a computation until it's required. The idea of lazy or non-strict evaluation is very helpful. To an extent, Python offers this feature.

In Python, the logical expression operators `and`, `or`, and `if-then-else` are all non-strict. We sometimes call them *short-circuit* operators because they don't need to evaluate all arguments to determine the resulting value.

The following command snippet shows the `and` operator's non-strict feature:

```
>>> 0 and print("right")
0
>>> True and print("right")
right
```

When we execute the first of the preceding command snippet, the left-hand side of the `and` operator is equivalent to `False`; the right-hand side is not evaluated. In the second example, when the left-hand side is equivalent to `True`, the right-hand side is evaluated.

Other parts of Python are strict. Outside the logical operators, an expression is evaluated eagerly from left to right. A sequence of statement lines is also evaluated strictly in order. Literal lists and `tuples` require eager evaluation.

When a class is created, the method functions are defined in a strict order. In the case of a class definition, the method functions are collected into a dictionary (by default) and order is not maintained after they're created. If we provide two methods with the same name, the second one is retained because of the strict evaluation order.

Python's generator expressions and generator functions, however, are lazy. These expressions don't create all possible results immediately. It's difficult to see this without explicitly logging the details of a calculation. Here is an example of the version of the `range()` function that has the side effect of showing the numbers it creates:

```
def numbers():
    for i in range(1024):
        print(f"= {i}")
        yield i
```

To provide some debugging hints, this function prints each value as the value is yielded. If this function were eager, it would create all 1,024 numbers. Since it's lazy, it only creates numbers as requested.

The older Python 2 `range()` function was eager and created an actual list object with all of the requested numbers. The Python 3 `range()` object is lazy, and will not create a large data structure.

We can use this noisy `numbers()` function in a way that will show lazy evaluation. We'll write a function that evaluates some, but not all, of the values from this iterator:

```
def sum_to(n: int) -> int:
    sum: int = 0
    for i in numbers():
        if i == n: break
        sum += i
    return sum
```

The `sum_to()` function has type hints to show that it should accept an integer value for the `n` parameter and return an integer result. The `sum` variable also includes Python 3 syntax, `: int`, a hint that it should be considered to be an integer. This function will not evaluate the entire result of the `numbers()` function. It will break after only consuming a few values from the `numbers()` function. We can see this consumption of values in the following log:

```
>>> sum_to(5)
= 0
= 1
= 2
= 3
= 4
= 5
10
```

As we'll see later, Python generator functions have some properties that make them a little awkward for simple functional programming. Specifically, a generator can only be used once in Python. We have to be cautious with how we use the lazy Python generator expressions.

Recursion instead of an explicit loop state

Functional programs don't rely on loops and the associated overhead of tracking the state of loops. Instead, functional programs try to rely on the much simpler approach of recursive functions. In some languages, the programs are written as recursions, but **Tail-CallOptimization** (**TCO**) in the compiler changes them to loops. We'll introduce some recursion here and examine it closely in `Chapter 6`, *Recursions and Reductions*.

We'll look at a simple iteration to test a number for being prime. A prime number is a natural number, evenly divisible by only 1 and itself. We can create a naïve and poorly-performing algorithm to determine whether a number has any factors between 2 and the number. This algorithm has the advantage of simplicity; it works acceptably for solving **Project Euler** problems. Read up on **Miller-Rabin** primality tests for a much better algorithm.

We'll use the term `coprime` to mean that two numbers have only one as their common factor. The numbers 2 and 3, for example, are `coprime`. The numbers 6 and 9, however, are not `coprime` because they have three as a common factor.

If we want to know whether a number, n, is prime, we actually ask this: is the number n `coprime` to all prime numbers, p, such that $p^2 < n$? We can simplify this using all integers, p, so that $2 \le p^2 < n$.

Sometimes, it helps to formalize this as follows:

$$\text{prime}(n) = \forall x[(2 \le x < 1 + \sqrt{n}) \wedge (n \mod x) \ne 0)]$$

The expression could look as follows in Python:

```
not any(n%p==0 for p in range(2,int(math.sqrt(n))+1))
```

A more direct conversion from mathematical formalism to Python would use `all(n%p != 0... )`, but that requires strict evaluation of all values of p. The *not any* version can terminate early if a `True` value is found.

This simple expression has a `for` loop inside it: it's not a pure example of stateless functional programming. We can reframe this into a function that works with a collection of values. We can ask whether the number, n, is `coprime` within any value in the half-open interval $[2, 1 + \sqrt{n})$. This uses the symbols `[)` to show a half-open interval: the lower values are included, and the upper value is not included. This is typical behavior of the Python `range()` function. We will also restrict ourselves to the domain of natural numbers. The square root values, for example, are implicitly truncated to integers.

We can think of the definition of prime as the following:

$\text{prime}(n) = \neg\text{coprime}(n, [2, 1 + \sqrt{n}))$, given `n > 1`.

When defining a recursive function over a simple range of values, the base case can be an empty range. A non-empty range is handled recursively by processing one value combined with a range that's narrower by one value. We could formalize it as follows:

$$\text{coprime}(n, [a, b)) = \begin{cases} \text{True} & \textbf{if } a = b \\ n \mod a \neq 0 \land \text{coprime}(n, [a+1, b)) & \textbf{if } a < b \end{cases}$$

This version is relatively easy to confirm by examining the two cases, which are given as follows:

- If the range is empty, $a = b$, we evaluated something like: $\text{coprime}(131073, [363, 363))$. The range contains no values, so the return is a trivial `True`.
- If the range is not empty, we evaluated something like: $\text{coprime}(131073, [2, 363))$. This decomposes into $131071 \mod 2 \neq 0 \land \text{coprime}(131071, [3, 363))$. For this example, we can see that the first clause is `True`, and we'll evaluate the second clause recursively.

As an exercise for the reader, this recursion can be redefined to count down instead of up, using `[a,b-1)` in the second case. Try this revision to see what, if any, changes are required.

As a side note, some folks like to think of the empty interval as $a \geq b$ instead of the $a = b$. The extra condition is needless, since a is incremented by one and we can easily guarantee that $a \leq b$, initially. There's no way for a to somehow leap past b through some error in the function; we don't need to over-specify the rules for an empty interval.

Here is a Python code snippet that implements this definition of prime:

```
def isprimer(n: int) -> bool:
    def isprime(k: int, coprime: int) -> bool:
        """Is k relatively prime to the value coprime?"""
        if k < coprime*coprime: return True
        if k % coprime == 0: return False
        return isprime(k, coprime+2)
    if n < 2: return False
    if n == 2: return True
    if n % 2 == 0: return False
    return isprime(n, 3)
```

This shows a recursive definition of an `isprimer()` function. The function expects an `int` value for the `n` parameter. The type hints suggest it will return a `bool` result.

The half-open interval, $(2, 1+\sqrt{n})$, is reduced to just the low-end argument, *a*, which is renamed to the `coprime` parameter to clarify its purpose. The base case is implemented as `n < coprime*coprime`; the range of values from `coprime` to `1+math.sqrt(n)` would be empty.

Rather than use a non-strict `and` operation, this example splits the decision into a separate `if` statement, `if n % coprime == 0`. The final `return` statement is the recursive call with a different `coprime` test value.

Because the recursion is the tail end of the function, this is an example of **Tail recursion**.

This `isprime()` function is embedded in the `isprimer()` function. The outer function serves to establish the boundary condition that *n* is an odd number greater than 2. There's no point in testing even numbers for being prime, since 2 is the only even prime.

What's important in this example is that the two cases of this recursive function are quite simple to design. Making the range of values an explicit argument to the internal `isprime()` function allows us to call the function recursively with argument values that reflect a steadily shrinking interval.

While recursion is often succinct and expressive, we have to be cautious about using it in Python. There are two problems that can arise:

- Python imposes a recursion limit to detect recursive functions with improperly defined base cases
- Python does not have a compiler that does TCO

The default recursion limit is 1,000, which is adequate for many algorithms. It's possible to change this with the `sys.setrecursionlimit()` function. It's not wise to raise this arbitrarily since it might lead to exceeding the OS memory limitations and crashing the Python interpreter.

If we try a recursive `isprimer()` function on a value of *n* over 1,000,000, we'll run afoul of the recursion limit. Even if we modified the `isprimer()` function to only check prime factors instead of all factors, we'd be stopped at the 1,000th prime number, 7,919, limiting our prime testing to numbers below 62,710,561.

Some functional programming languages can optimize simple recursive functions. An optimizing compiler will transform the recursive evaluation of the `isprimer(n, coprime+1)` method into a low-overhead `loop`. The optimization tends to make debugging optimized programs more difficult. Python doesn't perform this optimization. Performance and memory are sacrificed for clarity and simplicity. This also means we are forced to do the optimization manually.

In Python, when we use a generator expression instead of a recursive function, we essentially do the TCO manually. We don't rely on a compiler to do this optimization.

Here is TCO done as a generator expression:

```
def isprimei(n: int) -> bool:
    """Is n prime?

    >>> isprimei(2)
    True
    >>> tuple( isprimei(x) for x in range(3,11) )
    (True, False, True, False, True, False, False, False)
    """
    if n < 2:
        return False
    if n == 2:
        return True
    if n % 2 == 0:
        return False
    for i in range(3, 1+int(math.sqrt(n)), 2):
        if n % i == 0:
            return False
    return True
```

This function includes many of the functional programming principles, but it uses a generator expression instead of a pure recursion.

We'll often optimize a purely recursive function to use an explicit `for` loop in a generator expression.

This algorithm is slow for large primes. For composite numbers, the function often returns a value quickly. If used on a value such as $M_{61} = 2^{61} - 1$, it will take a few minutes to show that this is prime. Clearly, the slowness comes from checking 1,518,500,249 individual candidate factors.

Functional type systems

Some functional programming languages, such as **Haskell** and **Scala**, are statically compiled, and depend on declared types for functions and their arguments. To provide the kind of flexibility Python already has, these languages have sophisticated type-matching rules so that a generic function can be written, which works for a variety of related types.

In Object-Oriented Python, we often use the class inheritance hierarchy instead of sophisticated function type matching. We rely on Python to dispatch an operator to a proper method based on simple name-matching rules.

Since Python already has the desired levels of flexibility, the type matching rules for a compiled functional language aren't relevant. Indeed, we could argue that the sophisticated type matching is a workaround imposed by static compilation. Python doesn't need this workaround because it's a dynamic language.

Python 3 introduces type hints. These can be used by a program like **mypy** to discern potential problems with type mismatches. Using type hints is superior to using tests such ase `assert isinstance(a, int)` to detect whether an argument value for the a parameter is an `int`. An `assert` statement is a runtime burden. Running **mypy** to validate the hints is generally part of ordinary quality assurance. It's a common practice to run **mypy** and **pylint** along with unit tests to confirm that software is correct.

Familiar territory

One of the ideas that emerged from the previous list of topics is that most functional programming is already present in Python. Indeed, most functional programming is already a very typical and common part of OOP.

As a very specific example, a fluent **Application Program Interface** (**API**) is a very clear example of functional programming. If we take time to create a class with `return self()` in each method function, we can use it as follows:

```
some_object.foo().bar().yet_more()
```

We can just as easily write several closely related functions that work as follows:

```
yet_more(bar(foo(some_object)))
```

We've switched the syntax from traditional object-oriented suffix notation to a more functional prefix notation. Python uses both notations freely, often using a prefix version of a special method name. For example, the `len()` function is generally implemented by the `__len__()` class special method.

Of course, the implementation of the preceding class might involve a highly stateful object. Even then, a small change in viewpoint may reveal a functional approach that can lead to more succinct or more expressive programming.

The point is not that imperative programming is broken in some way, or that functional programming offers a vastly superior technology. The point is that functional programming leads to a change in viewpoint that can, in many cases, be helpful for designing succinct, expressive programs.

Learning some advanced concepts

We will set some more advanced concepts aside for consideration in later chapters. These concepts are part of the implementation of a purely functional language. Since Python isn't purely functional, our hybrid approach won't require deep consideration of these topics.

We will identify these up front for the benefit of folks who already know a functional language such as Haskell and are learning Python. The underlying concerns are present in all programming languages but we'll tackle them differently in Python. In many cases, we can and will drop into imperative programming rather than use a strictly functional approach.

The topics are as follows:

- **Referential transparency**: When looking at lazy evaluation and the various kinds of optimizations that are possible in a compiled language, the idea of multiple routes to the same object is important. In Python, this isn't as important because there aren't any relevant compile-time optimizations.
- **Currying**: The type systems will employ currying to reduce multiple-argument functions to single-argument functions. We'll look at currying in some depth in `Chapter 11`, *Decorator Design Techniques*.

- **Monads**: These are purely functional constructs that allow us to structure a sequential pipeline of processing in a flexible way. In some cases, we'll resort to imperative Python to achieve the same end. We'll also leverage the elegant `PyMonad` library for this. We'll defer this until `Chapter 14`, *The PyMonad Library*.

Summary

In this chapter, we've identified a number of features that characterize the functional programming paradigm. We started with first-class and higher-order functions. The idea is that a function can be an argument to a function or the result of a function. When functions become the object of additional programming, we can write some extremely flexible and generic algorithms.

The idea of immutable data is sometimes odd in an imperative and object-oriented programming language such as Python. When we start to focus on functional programming, however, we see a number of ways that state changes can be confusing or unhelpful. Using immutable objects can be a helpful simplification.

Python focuses on strict evaluation: all sub-expressions are evaluated from left-to-right through the statement. Python, however, does perform some non-strict evaluation. The `or`, `and`, and `if-else` logical operators are non-strict: all subexpressions are not necessarily evaluated. Similarly, a generator function is also non-strict. We can also call this eager vs. lazy. Python is generally eager, but we can leverage generator functions to create lazy evaluation.

While functional programming relies on recursion instead of the explicit loop state, Python imposes some limitations here. Because of the stack limitation and the lack of an optimizing compiler, we're forced to manually optimize recursive functions. We'll return to this topic in `Chapter 6`, *Recursions and Reductions*.

Although many functional languages have sophisticated type systems, we'll rely on Python's dynamic type resolution. In some cases, this means we'll have to write manual coercion for various types. It might also mean that we'll have to create class definitions to handle very complex situations. For the most part, however, Python's built-in rules will work very elegantly.

In the next chapter, we'll look at the core concepts of pure functions and how these fit in with Python's built-in data structures. Given this foundation, we can look at the higher-order functions available in Python and how we can define our own higher-order functions.

3
Functions, Iterators, and Generators

The core of functional programming is the use of pure functions to map values from the input domain to the output range. A pure function has no side effects, a relatively easy threshold for us to achieve in Python.

Avoiding side effects can lead to reducing any dependence on variable assignment to maintain the state of our computations. We can't purge the assignment statement from the Python language, but we can reduce our dependence on stateful objects. This means choosing among the available Python built-in functions and data structures to select those that don't require stateful operations.

This chapter will present several Python features from a functional viewpoint, as follows:

- Pure functions, free of side effects
- Functions as objects that can be passed as arguments or returned as results
- The use of Python strings using object-oriented suffix notation and prefix notation
- Using tuples and named tuples as a way to create stateless objects
- Using iterable collections as our primary design tool for functional programming

We'll look at generators and generator expressions, since these are ways to work with collections of objects. As we noted in `Chapter 2`, *Introducing Essential Functional Concepts*, there are some boundary issues when trying to replace all generator expressions with recursions. Python imposes a recursion limit, and doesn't automatically handle **Tail Call Optimization** (**TCO**): we must optimize recursions manually using a generator expression.

We'll write generator expressions that will perform the following tasks:

- Conversions
- Restructuring
- Complex calculations

We'll take a quick survey of many of the built-in Python collections, and how we can work with collections while pursuing a functional paradigm. This may change our approach to working with `lists`, `dicts`, and `sets`. Writing functional Python encourages us to focus on tuples and immutable collections. In the next chapter, we'll emphasize more functional ways to work with specific kinds of collections.

Writing pure functions

A function with no side effects fits the pure mathematical abstraction of a function: there are no global changes to variables. If we avoid the `global` statement, we will almost meet this threshold. To be **pure**, a function should also avoid changing the state mutable objects.

Here's an example of a pure function:

```
def m(n: int) -> int:
    return 2**n-1
```

This result depends only on the parameter, *n*. There are no changes to global variables and the function doesn't update any mutable data structures.

Any references to values in the Python global namespace (using a free variable) is something we can rework into a proper parameter. In most cases, it's quite easy. Here is an example that depends on a free variable:

```
def some_function(a: float, b: float, t: float) -> float:
    return a+b*t+global_adjustment
```

We can refactor this function to turn the `global_adjustment` variable into a proper parameter. We would need to change each reference to this function, which may have a large ripple effect through a complex application. A function with global references will include free variables in the body of a function.

There are many internal Python objects that are stateful. Instances of the `file` class and other file-like objects, are examples of stateful objects in common use. We observe that some of the commonly used stateful objects in Python generally behave as context managers. In a few cases, stateful objects don't completely implement the context manager interface; in these cases, there's often a `close()` method. We can use the `contextlib.closing()` function to provide these objects with the proper context manager interface.

We can't easily eliminate all stateful Python objects. Therefore, we must strike a balance between managing state while still exploiting the strengths of functional design. Toward this end, we should always use the `with` statement to encapsulate stateful file objects into a well-defined scope.

Always use file objects in a `with` context.

We should always avoid global file objects, global database connections, and the associated stateful object issues. The global file object is a common pattern for handling open files. We may have a function as shown in the following command snippet:

```
def open(iname: str, oname: str):
    global ifile, ofile
    ifile= open(iname, "r")
    ofile= open(oname, "w")
```

Given this context, numerous other functions can use the `ifile` and `ofile` variables, hoping they properly refer to the `global` files, which are left open for the application to use.

This is not a very functional design, and we need to avoid it. The files should be proper parameters to functions, and the open files should be nested in a `with` statement to assure that their stateful behavior is handled properly. This is an important rewrite to change these variables from globals to formal parameters: it makes the file operations more visible.

This design pattern also applies to databases. A database connection object should generally be provided as a formal argument to an application's functions. This is contrary to the way some popular web frameworks work: some frameworks rely on a global database connection in an effort to make the database a transparent feature of the application. This transparency obscures a dependency between a web operation and the database; it can make unit testing more complex than necessary. Additionally, a multithreaded web server may not benefit from sharing a single database connection: a connection pool is often better. This suggests that there are some benefits of a hybrid approach that uses functional design with a few isolated stateful features.

Functions as first-class objects

It shouldn't come as a surprise that Python functions are first-class objects. In Python, function objects have a number of attributes. The reference manual lists a number of special member names that apply to functions. Since functions are objects with attributes, we can extract the docstring or the name of a function, using special attributes such as `__doc__` or `__name__`. We can also extract the body of the function through the `__code__` attribute. In compiled languages, this introspection is relatively complex because of the source information that needs to be retained. In Python, it's quite simple.

We can assign functions to variables, pass functions as arguments, and return functions as values. We can easily use these techniques to write higher-order functions.

Additionally, a callable object helps us to create functions. We can consider the callable class definition as a higher-order function. We do need to be judicious in how we use the `__init__()` method of a callable object; we should avoid setting stateful class variables. One common application is to use an `__init__()` method to create objects that fit the `Strategy` design pattern.

A class following the `Strategy` design pattern depends on other objects to provide an algorithm or parts of an algorithm. This allows us to inject algorithmic details at runtime, rather than compiling the details into the class.

Here is an example of a callable object with an embedded `Strategy` object:

```
from typing import Callable
class Mersenne1:
    def __init__(self, algorithm : Callable[[int], int]) -> None:
        self.pow2 = algorithm
    def __call__(self, arg: int) -> int:
        return self.pow2(arg)-1
```

This class uses `__init__()` to save a reference to another function, `algorithm`, as `self.pow2`. We're not creating any stateful instance variables; the value of `self.pow2` isn't expected to change. The `algorithm` parameter has a type hint of `Callable[[int], int]`, a function that takes an integer argument and returns an integer value.

The function given as a `Strategy` object must raise 2 to the given power. Three candidate objects that we can plug into this class are as follows:

```
def shifty(b: int) -> int:
    return 1 << b

def multy(b: int) -> int:
    if b == 0: return 1
    return 2*multy(b-1)

def faster(b: int) -> int:
    if b == 0: return 1
    if b%2 == 1: return 2*faster(b-1)
    t= faster(b//2)
    return t*t
```

The `shifty()` function raises 2 to the desired power using a left shift of the bits. The `multy()` function uses a naive recursive multiplication. The `faster()` function uses a divide and conquer strategy that will perform $\log_2(b)$ multiplications instead of b multiplications.

All three of these functions have identical function signatures. Each of them can be summarized as `Callable[[int], int]`, which matches the parameter, `algorithm`, of the `Mersenne1.__init__()` method.

We can create instances of our `Mersenne1` class with an embedded strategy algorithm, as follows:

```
m1s = Mersenne1(shifty)
m1m = Mersenne1(multy)
m1f = Mersenne1(faster)
```

This shows how we can define alternative functions that produce the same result but use different algorithms.

Python allows us to compute $M_{89} = 2^{89} - 1$, since this doesn't even come close to the recursion limits in Python. This is quite a large prime number, as it has 27 digits.

Using strings

Since Python strings are immutable, they're an excellent example of functional programming objects. A Python `str` object has a number of methods, all of which produce a new string as the result. These methods are pure functions with no side effects.

The syntax for `str` method functions is postfix, where most functions are prefix. This means that complex string operations can be hard to read when they're co-mingled with conventional functions. For example, in this expression, `len(variable.title())`, the `title()` method is in postfix notation and the `len()` function is in prefix notation.

When scraping data from a web page, we may have a function to clean the data. This could apply a number of transformations to a string to clean up the punctuation and return a `Decimal` object for use by the rest of the application. This will involve a mixture of prefix and postfix syntax.

It could look like the following command snippet:

```
from decimal import *
from typing import Text, Optional
def clean_decimal(text: Text) -> Optional[Text]:
    if text is None: return None
    return Decimal(
        text.replace("$", "").replace(",", ""))
```

This function does two replacements on the string to remove `$` and `,` string values. The resulting string is used as an argument to the `Decimal` class constructor, which returns the desired object. If the input value is `None`, this is preserved; this is why the `Optional` type hint is used.

To make the syntax look more consistent, we can consider defining our own prefix functions for the `string` method functions, as follows:

```
def replace(str: Text, a: Text, b: Text) -> Text:
    return str.replace(a,b)
```

This can allow us to use `Decimal(replace(replace(text, "$", ""), ",", ""))` with consistent-looking prefix syntax. It's not clear whether this kind of consistency is a significant improvement over the mixed prefix and postfix notation. This may be an example of a foolish consistency.

A slightly better approach may be to define a more meaningful prefix function to strip punctuation, such as the following command snippet:

```
def remove(str: Text, chars: Text) -> Text:
    if chars:
        return remove(
            str.replace(chars[0], ""),
            chars[1:]
        )
    return str
```

This function will recursively remove each of the characters from the `chars` variable. We can use it as `Decimal(remove(text, "$,"))` to make the intent of our string cleanup more clear.

Using tuples and named tuples

Since Python tuples are immutable objects, they're another excellent example of objects suitable for functional programming. A Python `tuple` has very few method functions, so almost everything is done using prefix syntax. There are a number of use cases for tuples, particularly when working with list-of-tuple, tuple-of-tuple, and generator-of-tuple constructs.

The `namedtuple` class adds an essential feature to a tuple: a name that we can use instead of an index. We can exploit named tuples to create objects that are accretions of data. This allows us to write pure functions based on stateless objects, yet keep data bound into tidy object-like packages.

We'll almost always use tuples (and named tuples) in the context of a collection of values. If we're working with single values, or a tidy group of exactly two values, we'll usually use named parameters to a function. When working with collections, however, we may need to have iterable-of-tuples or iterable of the `namedtuple` class constructs.

The decision to use a `tuple` or `namedtuple` object is entirely a matter of convenience. As an example, consider working with a sequence of color values as a three tuple of the form `(number, number, number)` It's not clear that these are in red, green, blue order. We have a number of approaches for making the tuple structure explicit.

We can clarify the triple structure by creating functions to pick a three-tuple apart, as shown in the following command snippet:

```
red = lambda color: color[0]
green = lambda color: color[1]
blue = lambda color: color[2]
```

Given a tuple, `item`, we can use `red(item)` to select the item that has the red component. It can help to provide a more formal type hint on each variable, as follows:

```
from typing import Tuple, Callable
RGB = Tuple[int, int, int]
red: Callable[[RGB], int] = lambda color: color[0]
```

This defines a new type, `RGB`, as a three-tuple. The `red` variable is provided with a type hint of `Callable[[RGB], int]` to indicate it should be considered to be a function that accepts an `RGB` argument and produces an integer result.

Or, we may introduce the following definition using older-style `namedtuple` class objects:

```
from collections import namedtuple
Color = namedtuple("Color", ("red", "green", "blue", "name"))
```

An even better technique is to use the `NamedTuple` class from the typing module:

```
from typing import NamedTuple
class Color(NamedTuple):
    """An RGB color."""
    red: int
    green: int
    blue: int
    name: str
```

This definition of the `Color` class defines a tuple with specific names and type hints for each position within the tuple. This preserves the advantages of performance and immutability. It adds the ability for the **mypy** program to confirm that the tuple is used properly.

These three technicals allow the use of `item.red` instead of `red(item)`. Either of these are better than the confusing `item[0]`.

The functional programming application of tuples centers on the iterable-of-tuple design pattern. We'll look closely at a few iterable-of-tuple techniques. We'll look at the `namedtuple` class techniques in `Chapter 7`, *Additional Tuple Techniques*.

Using generator expressions

We've shown some examples of generator expressions already. We'll show some more later in the chapter. We'll introduce some more generator techniques in this section.

It's common to see generator expressions used to create the `list` or `dict` literals through `list` comprehension or a `dict` comprehension syntax. This example, `[x**2 for x in range(10)]`, is a list comprehension, sometimes called a list display. For our purposes, the list display (or comprehension) is one of several ways to use generator expressions. A collection display includes the enclosing literal syntax. In this example, the list literal `[]` characters wrap the generator: `[x**2 for x in range(10)]`. This is a list comprehension; it creates a `list` object from the enclosed generator expression, `x**2 for x in range(10)`. In this section, we're going to focus on the generator expression separate from the list object.

A collection object and a generator expression have some similar behaviors because both are iterable. They're not equivalent, as we'll see in the following code. Using displays has the disadvantage of creating a (potentially large) collection of objects. A generator expression is lazy and creates objects only as required; this can improve performance.

We have to provide two important caveats on generator expressions, as follows:

- Generators appear to be sequence-like. The few exceptions include using a function such as the `len()` function that needs to know the size of the collection.
- Generators can be used only once. After that, they appear empty.

Here is a generator function that we'll use for some examples:

```
def pfactorsl(x: int) -> Iterator[int]:
    if x % 2 == 0:
        yield 2
        if x//2 > 1:
            yield from pfactorsl(x//2)
        return
    for i in range(3, int(math.sqrt(x)+.5)+1, 2):
        if x % i == 0:
            yield i
            if x//i > 1:
                yield from pfactorsl(x//i)
            return
    yield x
```

We're locating the prime factors of a number. If the number, *x*, is even, we'll yield 2 and then recursively yield all prime factors of *x*/2.

For odd numbers, we'll step through odd values greater than or equal to 3 to locate a candidate factor of the number. When we locate a factor, *i*, we'll yield that factor, and then recursively yield all prime factors of $x \div i$.

In the event that we can't locate a factor, the number, *x*, must be prime, so we can yield the number.

We handle 2 as a special case to cut the number of iterations in half. All prime numbers, except 2, are odd.

We've used one important `for` loop in addition to recursion. This explicit loop allows us to easily handle numbers that have as many as 1,000 factors. (As an example, $2^{1,000}$, a number with 300 digits, will have 1,000 factors.) Since the `for` variable, `i`, is not used outside the indented body of the loop, the stateful nature of the `i` variable won't lead to confusion if we make any changes to the body of the loop.

This example shows how to do tail-call optimization manually. The recursive calls that count from 3 to $\sqrt{x}$ have been replaced with a loop. The `for` loop saves us from a deeply recursive call stack.

Because the function is iterable, the `yield from` statement is used to consume iterable values from the recursive call and yield them to the caller.

In a recursive generator function, be careful of the return statement. Do not use the following command line: `return recursive_iter(args)`
It returns only a generator object; it doesn't evaluate the function to return the generated values. Use any of the following:

```
for result in recursive_iter(args):
    yield result
yield from recursive_iter(args)
```

As an alternative, the following definition is a more purely recursive version:

```
def pfactorsr(x: int) -> Iterator[int]:
    def factor_n(x: int, n: int) -> Iterator[int]:
        if n*n > x:
            yield x
            return
        if x % n == 0:
            yield n
            if x//n > 1:
                yield from factor_n(x//n, n)
        else:
            yield from factor_n(x, n+2)
```

```
    if x % 2 == 0:
        yield 2
        if x//2 > 1:
            yield from pfactorsr(x//2)
        return
    yield from factor_n(x, 3)
```

We defined an internal recursive function, `factor_n()`, to test factors, *n*, in the range $3 \leq n \leq \sqrt{x}$. If the candidate factor, *n*, is outside the range, then *x* is prime. Otherwise, we'll see whether *n* is a factor of *x*. If so, we'll yield *n* and all factors of $\frac{x}{n}$. If *n* is not a factor, we'll evaluate the function recursively using $n+2$. This uses recursion to test each value of $(n+2, n+2+2, n+2+2, \ldots)$. While this is simpler than the `for` statement version shown previously, it can't handle numbers with over 1,000 factors because of Python's stack limitation.

The outer function handles some edge cases. As with other prime-related processing, we handle two as a special case. For even numbers, we'll yield two and then evaluate `pfactorsr()` recursively for $x \div 2$. All other prime factors must be odd numbers greater than or equal to 3. We'll evaluate the `factors_n()` function starting with 3 to test these other candidate prime factors.

The purely recursive function can only locate prime factors of numbers up to about 4,000,000. Above this, Python's recursion limit will be reached.

Exploring the limitations of generators

We noted that there are some limitations of generator expressions and generator functions. The limitations can be observed by executing the following command snippet:

```
>>> from ch02_ex4 import *
>>> pfactorsl(1560)
<generator object pfactorsl at 0x1007b74b0>
>>> list(pfactorsl(1560))
[2, 2, 2, 3, 5, 13]
>>> len(pfactorsl(1560))
Traceback (most recent call last):
  File "<stdin>", line 1, in <module>
TypeError: object of type 'generator' has no len()
```

In the first example, we saw the generator function, `pfactors1`, created a generator. The generator is lazy, and doesn't have a proper value until we consume the results yielded by the generator. in itself isn't a limitation; lazy evaluation is an important reason why generator expressions fit with functional programming in Python.

In the second example, we materialized a `list` object from the results yielded by the generator function. This is handy for seeing the output and writing unit test cases.

In the third example, we saw one limitation of generator functions: there's no `len()`. Because the generator is lazy, the size can't be known until after all of the values are consumed.

The other limitation of generator functions is that they can only be used once.

For example, look at the following command snippet:

```
>>> result = pfactorsl(1560)
>>> sum(result)
27
>>> sum(result)
0
```

The first evaluation of the `sum()` method performed evaluation of the generator, `result`. All of the values were consumed. The second evaluation of the `sum()` method found that the generator was now empty. We can only consume the values of a generator once.

Generators have a stateful life in Python. While they're very nice for some aspects of functional programming, they're not quite perfect.

We can try to use the `itertools.tee()` method to overcome the once-only limitation. We'll look at this in depth in `Chapter 8`, *The Itertools Module*. Here is a quick example of its usage:

```
import itertools
from typing import Iterable, Any
def limits(iterable: Iterable[Any]) -> Any:
    max_tee, min_tee = itertools.tee(iterable, 2)
    return max(max_tee), min(min_tee)
```

We created two clones of the parameter generator expression, `max_tee` and `min_tee`. This leaves the original iterator untouched, a pleasant feature that allows us to do very flexible combinations of functions. We can consume these two clones to get maximum and minimum values from the iterable.

Once consumed, an iterable will not provide any more values. When we want to compute multiple kinds of reductions—for example, sums and counts, or minimums and maximums—we need to design with this one-pass-only limitation in mind.

Combining generator expressions

The essence of functional programming comes from the ways we can easily combine generator expressions and generator functions to create very sophisticated composite processing sequences. When working with generator expressions, we can combine generators in several ways.

One common way to combine generator functions is when we create a composite function. We may have a generator that computes `(f(x) for x in range())`. If we want to compute `g(f(x))`, we have several ways to combine two generators.

We can tweak the original generator expression as follows:

```
g_f_x = (g(f(x)) for x in range())
```

While technically correct, this defeats any idea of reuse. Rather than reusing an expression, we rewrote it.

We can also substitute one expression within another expression, as follows:

```
g_f_x = (g(y) for y in (f(x) for x in range()))
```

This has the advantage of allowing us to use simple substitution. We can revise this slightly to emphasize reuse, using the following commands:

```
f_x = (f(x) for x in range())
g_f_x = (g(y) for y in f_x)
```

This has the advantage of leaving the initial expression, `(f(x) for x in range())`, essentially untouched. All we did was assign the expression to a variable.

The resulting composite function is also a generator expression, which is also lazy. This means that extracting the next value from `g_f_x` will extract one value from `f_x`, which will extract one value from the source `range()` function.

Cleaning raw data with generator functions

One of the tasks that arise in exploratory data analysis is cleaning up raw source data. This is often done as a composite operation applying several scalar functions to each piece of input data to create a usable dataset.

Let's look at a simplified set of data. This data is commonly used to show techniques in exploratory data analysis. It's called `Anscombe's quartet`, and it comes from the article, *Graphs in Statistical Analysis*, by F. J. Anscombe that appeared in *American Statistician* in 1973. The following are the first few rows of a downloaded file with this dataset:

```
Anscombe's quartet
I  II  III  IV
x  y  x  y  x  y  x  y
10.0  8.04  10.0  9.14      10.0  7.46  8.0  6.58
8.0       6.95  8.0  8.14  8.0  6.77  8.0  5.76
13.0  7.58  13.0  8.74  13.0  12.74  8.0  7.71
```

Sadly, we can't trivially process this with the `csv` module. We have to do a little bit of parsing to extract the useful information from this file. Since the data is properly tab-delimited, we can use the `csv.reader()` function to iterate through the various rows. We can define a data iterator as follows:

```
import csv
from typing import IO, Iterator, List, Text, Union, Iterable
def row_iter(source: IO) -> Iterator[List[Text]]:
    return csv.reader(source, delimiter="\t")
```

We simply wrapped a file in a `csv.reader` function to create an iterator over rows. The typing module provides a handy definition, `IO`, for file objects. The purpose of the `csv.reader()` function is to be an iterator over the rows. Each row is a list of text values. It can be helpful to define an additional type `Row = List[Text]`, to make this more explicit.

We can use this `row_iter()` function in the following context:

```
with open("Anscombe.txt") as source:
    print(list(row_iter(source)))
```

While this will display useful information, the problem is the first three items in the resulting iterable aren't data. The `Anscombe's quartet` file starts with the following rows:

```
[["Anscombe's quartet"],
 ['I', 'II', 'III', 'IV'],
 ['x', 'y', 'x', 'y', 'x', 'y', 'x', 'y'],
```

We need to filter these three non-data rows from the iterable. Here is a function that will neatly excise three expected title rows, and return an iterator over the remaining rows:

```
def head_split_fixed(
        row_iter: Iterator[List[Text]]
    ) -> Iterator[List[Text]]:
    title = next(row_iter)
    assert (len(title) == 1
        and title[0] == "Anscombe's quartet")
    heading = next(row_iter)
    assert (len(heading) == 4
        and heading == ['I', 'II', 'III', 'IV'])
    columns = next(row_iter)
    assert (len(columns) == 8
        and columns == ['x','y', 'x','y', 'x','y', 'x','y'])
    return row_iter
```

This function plucks three rows from the source data, an iterator. It asserts that each row has an expected value. If the file doesn't meet these basic expectations, it's a sign that the file was damaged or perhaps our analysis is focused on the wrong file.

Since both the `row_iter()` and the `head_split_fixed()` functions expect an iterator as an argument value, they can be trivially combined, as follows:

```
with open("Anscombe.txt") as source:
    print(list(head_split_fixed(row_iter(source))))
```

We've simply applied one iterator to the results of another iterator. In effect, this defines a composite function. We're not done of course; we still need to convert the `strings` values to the `float` values, and we also need to pick apart the four parallel series of data in each row.

The final conversions and data extractions are more easily done with higher-order functions, such as `map()` and `filter()`. We'll return to those in `Chapter 5`, *Higher-Order Functions*.

Using lists, dicts, and sets

A Python sequence object, such as a `list`, is iterable. However, it has some additional features. We'll think of it as a materialized iterable. We've used the `tuple()` function in several examples to collect the output of a generator expression or generator function into a single `tuple` object. We can also materialize a sequence to create a `list` object.

In Python, a *list display*, or *list comprehension*, offers simple syntax to materialize a generator: we just add the `[]` brackets. This is ubiquitous to the point where the distinction between generator expression and list comprehension is lost. We need to disentangle the idea of generator expression from a list display that uses a generator expression.

The following is an example to enumerate the cases:

```
>>> range(10)
range(0, 10)
>>> [range(10)]
[range(0, 10)]
>>> [x for x in range(10)]
[0, 1, 2, 3, 4, 5, 6, 7, 8, 9]
>>> list(range(10))
[0, 1, 2, 3, 4, 5, 6, 7, 8, 9]
```

The first example is the `range` object, which is a type of generator function. It doesn't produce any values because it's lazy.

The `range(10)` function is lazy; it won't produce the 10 values until evaluated in a context that iterates through the values.

The second example shows a list composed of a single instance of the generator function. The `[]` syntax created a list literal of the `range()` object without consuming any values created by the iterator.

The third example shows a `list` comprehension built from a generator expression that includes a generator function. The function, `range(10)`, is evaluated by a generator expression, `x for x in range(10)`. The resulting values are collected into a `list` object.

We can also use the `list()` function to build a list from an iterable or a generator expression. This also works for `set()`, `tuple()`, and `dict()`.

The `list(range(10))` function evaluated the generator expression. The `[range(10)]` list literal does not evaluate the generator function.

While there's shorthand syntax for `list`, `dict`, and `set` using `[]` and `{}`, there's no shorthand syntax for a tuple. To materialize a tuple, we must use the `tuple()` function. For this reason, it often seems most consistent to use the `list()`, `tuple()`, and `set()` functions as the preferred syntax.

In the data-cleansing code, we used a composite function to create a list of four tuples. The function looked as follows:

```
with open("Anscombe.txt") as source:
    data = head_split_fixed(row_iter(source))
    print(list(data))
```

We assigned the results of the composite function to a name, `data`. The data looks as follows:

```
[['10.0', '8.04', '10.0', '9.14', '10.0', '7.46', '8.0', '6.58'],
 ['8.0', '6.95', '8.0', '8.14', '8.0', '6.77', '8.0', '5.76'],
 ...
 ['5.0', '5.68', '5.0', '4.74', '5.0', '5.73', '8.0', '6.89']]
```

We need to do a little bit more processing to make this useful. First, we need to pick pairs of columns from the eight-tuple. We can select pair of columns with a function, as shown in the following command snippet:

```
from typing import Tuple, cast

Pair = Tuple[str, str]
def series(
        n: int, row_iter: Iterable[List[Text]]
    ) -> Iterator[Pair]:
    for row in row_iter:
        yield cast(Pair, tuple(row[n*2:n*2+2]))
```

This function picks two adjacent columns based on a number between 0 and 3. It creates a `tuple` object from those two columns. The `cast()` function is a type hint to inform the **mypy** tool that the result will be a two-tuple where both items are strings. This is required because it's difficult for the **mypy** tool to determine that the expression `tuple(row[n*2:n*2+2])` will select exactly two elements from the `row` collection.

We can now create a tuple-of-tuples collection, as follows:

```
with open("Anscombe.txt") as source:
    data = tuple(head_split_fixed(row_iter(source)))
    sample_I = tuple(series(0, data))
    sample_II = tuple(series(1, data))
    sample_III = tuple(series(2, data))
    sample_IV = tuple(series(3, data))
```

We applied the `tuple()` function to a composite function based on the `head_split_fixed()` and `row_iter()` methods. This will create an object that we can reuse in several other functions. If we don't materialize a `tuple` object, only the first sample will have any data. After that, the source iterator will be exhausted and all other attempts to access it would yield empty sequences.

The `series()` function will pick pairs of items to create the `Pair` objects. Again, we applied an overall `tuple()` function to materialize the resulting tuple-of-named tuple sequences so that we can do further processing on each one.

The `sample_I` sequence looks as follows:

```
(('10.0', '8.04'), ('8.0', '6.95'), ('13.0', '7.58'),
('9.0', '8.81'), ('11.0', '8.33'), ('14.0', '9.96'),
('6.0', '7.24'), ('4.0', '4.26'), ('12.0', '10.84'),
('7.0', '4.82'), ('5.0', '5.68'))
```

The other three sequences are similar in structure. The values, however, are quite different.

The final thing we'll need to do is create proper numeric values from the strings that we've accumulated so that we can compute some statistical summary values. We can apply the `float()` function conversion as the last step. There are many alternative places to apply the `float()` function, and we'll look at some choices in `Chapter 5`, *Higher Order Functions*.

Here is an example describing the usage of the `float()` function:

```
mean = (
    sum(float(pair[1]) for pair in sample_I) / len(sample_I)
)
```

This will provide the mean of the *y* value in each two-tuple. We can gather a number of statistics as follows:

```
for subset in sample_I, sample_II, sample_III, sample_III:
    mean = (
        sum(float(pair[1]) for pair in subset)/len(subset)
    )
    print(mean)
```

We computed a mean for the `y` values in each two-tuple built from the source database. We created a common tuple of the `namedtuple` class structure so that we can have reasonably clear references to members of the source dataset. Using `pair[1]` can be an obscure way to reference a data item. In `Chapter 7`, *Additional Tuple Techniques*, we'll use named tuples to simplify references to items within a complex tuple.

To reduce memory use-and increase performance we prefer to use generator expressions and functions as much as possible. These iterate through collections in a lazy (or non-strict) manner, computing values only when required. Since iterators can only be used once, we're sometimes forced to materialize a collection as a `tuple` (or `list`) object. Materializing a collection costs memory and time, so we do it reluctantly.

Programmers familiar with **Clojure** can match Python's lazy generators with the `lazy-seq` and `lazy-cat` functions. The idea is that we can specify a potentially infinite sequence, but only take values from it as needed.

Using stateful mappings

Python offers several stateful collections; the various mappings include the dict class and a number of related mappings defined in the `collections` module. We need to emphasize the stateful nature of these mappings and use them carefully.

For our purposes, learning functional programming techniques in Python, there are two use cases for `mapping`: a stateful dictionary that accumulates a mapping and a frozen dictionary. In the first example of this chapter, we showed a frozen dictionary that was used by the `ElementTree.findall()` method. Python doesn't provide an easy-to-use definition of an immutable mapping. The `collections.Mapping` abstract class is immutable, but it's not something we can use trivially. We'll dive into details in `Chapter 6`, *Recursions and Reductions*.

Instead of the formality of using the `collections.Mapping` abstract class, we can fall back on confirming that the variable `ns_map` appears exactly once on the left side of an assignment statement; methods such as `ns_map.update()` or `ns_map.pop()` are never used, and the `del` statement isn't used with map items.

The stateful dictionary can be further decomposed into two typical use cases; they are as follows:

- A dictionary built once and never updated. In this case, we will exploit the hashed keys feature of the `dict` class to optimize performance. We can create a dictionary from any iterable sequence of (key, value) two tuples through `dict(sequence)`.

- A dictionary built incrementally. This is an optimization we can use to avoid materializing and sorting a list object. We'll look at this in `Chapter 6`, *Recursions and Reductions*. We'll look at the `collections.Counter` class as a sophisticated reduction. Incremental building is particularly helpful for memoization. We'll defer memoization until `Chapter 16`, *Optimizations and Improvements*.

The first example, building a dictionary once, stems from an application with three operating phases: gather some input, create a `dict` object, and then process input based on the mappings in the dictionary. As an example of this kind of application, we may be doing some image processing and have a specific palette of colors, represented by names and (R, G, B) tuples. If we use the **GNU Image Manipulation Program** (**GIMP**) file format, the color palette may look like the following command snippet:

```
GIMP Palette
Name: Small
Columns: 3
#
  0   0   0    Black
255 255 255    White
238  32  77    Red
28  172 120    Green
31  117 254    Blue
```

The details of parsing this file are the subject of `Chapter 6`, *Recursions and Reductions*. What's important is the results of the parsing.

First, we'll use the `namedtuple` class `Color` as follows:

```
from collections import namedtuple
Color = namedtuple("Color", ("red", "green", "blue", "name"))
```

Second, we'll assume that we have a parser that produces an iterable of `Color` objects. If we materialize it as a tuple, it would look like the following:

```
(Color(red=239, green=222, blue=205, name='Almond'),
 Color(red=205, green=149, blue=117, name='Antique Brass'),
 Color(red=253, green=217, blue=181, name='Apricot'),
 Color(red=197, green=227, blue=132, name='Yellow Green'),
 Color(red=255, green=174, blue=66, name='Yellow Orange'))
```

To locate a given color name quickly, we will create a frozen dictionary from this sequence. This is not the only way to get fast lookups of a color by name. We'll look at another option later.

To create a mapping from a tuple, we will use the `process(wrap(iterable))` design pattern. The following command shows how we can create the color name mapping:

```
name_map = dict((c.name, c) for c in sequence)
```

Here, the sequence variable is the iterable of the `Color` objects shown previously; the `wrap()` element of the design pattern simply transforms each `Color` object , *c*, into the two tuple `(c.name, c)`. The `process()` element of the design uses `dict()` initialization to create a mapping from name to `Color`. The resulting dictionary looks as follows:

```
{'Caribbean Green': Color(red=28, green=211, blue=162,
 name='Caribbean Green'),
 'Peach': Color(red=255, green=207, blue=171, name='Peach'),
 'Blizzard Blue': Color(red=172, green=229, blue=238, name='Blizzard
 Blue'),
 etc.
}
```

The order is not guaranteed, so you may not see Caribbean Green first.

Now that we've materialized the mapping, we can use this `dict()` object in some later processing for repeated transformations from color name to (R, G, B) color numbers. The lookup will be blazingly fast because a dictionary does a rapid transformation from key to hash value followed by lookup in the dictionary.

Using the bisect module to create a mapping

In the previous example, we created a `dict` mapping to achieve a fast mapping from a color name to a `Color` object. This isn't the only choice; we can use the `bisect` module instead. Using the `bisect` module means that we have to create a sorted object, which we can then search. To be perfectly compatible with the `dict` mapping, we can use `collections.Mapping` as the base class.

The `dict` mapping uses a hash to locate items almost immediately. However, this requires allocating a fairly large block of memory. The `bisect` mapping does a search, which doesn't require as much memory, but performance can be described as immediate.

A `static` mapping class looks like the following command snippet:

```
import bisect
from collections import Mapping
from typing import Iterable, Tuple, Any
```

```
class StaticMapping(Mapping):
    def __init__(self,
      iterable: Iterable[Tuple[Any, Any]]) -> None:
        self._data = tuple(iterable)
        self._keys = tuple(sorted(key for key,_ in self._data))
    def __getitem__(self, key):
        ix= bisect.bisect_left(self._keys, key)
        if (ix != len(self._keys)
          and self._keys[ix] == key_:
            return self._data[ix][1]
        raise ValueError("{0!r} not found".format(key))
    def __iter__(self):
        return iter(self._keys)
    def __len__(self):
        return len(self._keys)
```

This class extends the abstract superclass `collections.Mapping`. It provides an initialization and implementations for three functions missing from the abstract definition. The type of `Tuple[Any, Any]` defines a generic two-tuple.

The `__getitem__()` method uses the `bisect.bisect_left()` function to search the collection of keys. If the key is found, the appropriate value is returned. The `__iter__()` method returns an iterator, as required by the superclass. The `__len__()` method, similarly, provides the required length of the collection.

Another option is to start with the source code for the `collections.OrderedDict` class, change the superclass to `Mapping` instead of `MutableMapping`, and remove all of the methods that implement mutability. For more details on which methods to keep and which to discard, refer to the *Python Standard Library*, section 8.4.1.

Visit the following link for more details:

`https://docs.python.org/3.3/library/collections.abc.html#collections-abstract-base-classes`

This class may not seem to embody too many functional programming principles. Our goal here is to support a larger application that minimizes the use of stateful variables. This class saves a static collection of key-value pairs. As an optimization, it materializes two objects.

An application would create an instance of this class to perform rapid lookups of values associated with keys. The superclass does not support updates to the object. The collection, as a whole, is stateless. It's not as fast as the built-in `dict` class, but it uses less memory and, through the formality of being a subclass of `Mapping`, we can be assured that this object is not used to contain a processing state.

Using stateful sets

Python offers several stateful collections, including the set collection. For our purposes, there are two use cases for a set: a stateful set that accumulates items, and `frozenset` that is used to optimize searches for an item.

We can create `frozenset` from an iterable in the same way we create a `tuple` object from an iterable `fronzenset(some_iterable)` method; this will create a structure that has the advantage of a very fast `in` operator. This can be used in an application that gathers data, creates a set, and then uses that `frozenset` to process some other data items.

We may have a set of colors that we will use as a kind of **chroma-key**: we will use this color to create a mask that will be used to combine two images. Pragmatically, a single color isn't appropriate, but a small set of very similar colors works best. In this case, we'll examine each pixel of an image file to see if the pixel is in the chroma-key set or not. For this kind of processing, the chroma-key colors are loaded into `frozenset` before processing the target images. For more information, read about chroma-key processing from the following link:

`http://en.wikipedia.org/wiki/Chroma_key`

As with mappings—specifically the `Counter` class—there are some algorithms that can benefit from a memoized set of values. Some functions benefit from memoization because a function is a mapping between domain values and range values, a job where mapping works well. A few algorithms benefit from a memoized set, which is stateful and grows as data is processed.

We'll return to memoization in `Chapter 16`, *Optimizations and Improvements*.

Summary

In this chapter, we looked closely at writing pure functions free of side effects. The bar is low here, since Python forces us to use the `global` statement to write impure functions. We looked at generator functions and how we can use these as the backbone of functional programming. We also examined the built-in collection classes to show how they're used in the functional paradigm. While the general idea behind functional programming is to limit the use of stateful variables, the collection objects are generally stateful and, for many algorithms, also essential. Our goal is to be judicious in our use of Python's non-functional features.

In the next two chapters, we'll look at higher-order functions: functions that accept functions as arguments as well as returning functions. We'll start with an exploration of the built-in higher-order functions. In later chapters, we'll look at techniques for defining our own higher-order functions. We'll also look at the `itertools` and `functools` modules and their higher-order functions in later chapters.

4 Working with Collections

Python offers a number of functions that process whole collections. They can be applied to sequences (lists or tuples), sets, mappings, and iterable results of generator expressions. We'll look at Python's collection-processing features from a functional programming viewpoint.

We'll start out by looking at iterables and some simple functions that work with iterables. We'll look at some design patterns to handle iterables and sequences with recursive functions as well as explicit `for` loops. We'll look at how we can apply a scalar function to a collection of data with a generator expression.

In this chapter, we'll show you examples of how to use the following functions with collections:

- `any()` and `all()`
- `len()`, `sum()`, and some higher-order statistical processing related to these functions
- `zip()` and some related techniques to structure and flatten lists of data
- `reversed()`
- `enumerate()`

The first four functions can be called **reductions**: they reduce a collection to a single value. The other three functions, `zip()`, `reversed()`, and `enumerate()`, are mappings; they produce a new collection from an existing collection(s). In the next chapter, we'll look at some more mapping and reduction functions that use an additional function as an argument to customize their processing.

In this chapter, we'll start out by looking at ways to process data, using generator expressions. Then, we'll apply different kinds of collection-level functions to show how they can simplify the syntax of iterative processing. We'll also look at some different ways of restructuring data.

In the next chapter, we'll focus on using higher-order collection functions to do similar kinds of processing.

An overview of function varieties

We need to distinguish between two broad species of functions, as follows:

- **Scalar functions**: They apply to individual values and compute an individual result. Functions such as `abs()`, `pow()`, and the entire `math` module are examples of scalar functions.
- **Collection functions**: They work with iterable collections.

We can further subdivide the collection functions into three subspecies:

- **Reduction**: This uses a function to fold values in the collection together, resulting in a single final value. For example, if we fold (+) operations into a sequence of integers, this will compute the sum. This can be also be called an **aggregate function**, as it produces a single aggregate value for an input collection.
- **Mapping**: This applies a scalar function to each individual item of a collection; the result is a collection of the same size.
- **Filter**: This applies a scalar function to all items of a collection to reject some items and pass others. The result is a subset of the input.

We'll use this conceptual framework to characterize ways in which we use the built-in collection functions.

Working with iterables

As noted in the previous chapters, Python's `for` loop works with collections. When working with materialized collections such as tuples, lists, maps, and sets, the `for` loop involves the explicit management of states. While this strays from purely functional programming, it reflects a necessary optimization for Python. If we assume that state management is localized to an iterator object that's created as a part of the `for` statement evaluation, we can leverage this feature without straying too far from pure, functional programming. If, for example, we use the `for` loop variable outside the indented body of the `loop`, we've strayed from purely functional programming by leveraging this state control variable.

We'll return to this in `Chapter 6`, *Recursions and Reductions*. It's an important topic, and we'll just scratch the surface here with a quick example of working with generators.

One common application of `for` loop iterable processing is the `unwrap(process(wrap(iterable)))` design pattern. A `wrap()` function will first transform each item of an iterable into a two-tuple with a derived sort key and the original item. We can then process these two-tuple items as a single, wrapped value. Finally, we'll use an `unwrap()` function to discard the value used to wrap, which recovers the original item.

This happens so often in a functional context that two functions are used heavily for this; they are the following:

```
fst = lambda x: x[0]
snd = lambda x: x[1]
```

These two functions pick the first and second values from a two-tuple, and both are handy for the `process()` and `unwrap()` functions.

Another common pattern is `wrap3(wrap2(wrap1()))`. In this case, we're starting with simple tuples and then wrapping them with additional results to build up larger and more complex tuples. A common variation on this theme builds new, more complex `namedtuple` instances from source objects. We can summarize both of these as the **Accretion design pattern**—an item that accretes derived values.

As an example, consider using the accretion pattern to work with a simple sequence of latitude and longitude values. The first step will convert the simple point represented as a (`lat, lon`) pair on a path into pairs of legs (`begin, end`). Each pair in the result will be represented as ((`lat, lon`), (`lat, lon`)). The value of `fst(item)` is the starting position; the value of `snd(item)` is the ending position for each value of each item in the collection.

In the next sections, we'll show you how to create a generator function that will iterate over the content of a file. This iterable will contain the raw input data that we will process.

Once we have the raw data, later sections will show how to decorate each leg with the `haversine` distance along the leg. The final result of a `wrap(wrap(iterable()))` design will be a sequence of three tuples—((`lat, lon`), (`lat, lon`), `distance`). We can then analyze the results for the longest and shortest distance, bounding rectangle, and other summaries.

Parsing an XML file

We'll start by parsing an **Extensible Markup Language** (**XML**) file to get the raw latitude and longitude pairs. This will show you how we can encapsulate some not-quite-functional features of Python to create an iterable sequence of values.

We'll make use of the `xml.etree` module. After parsing, the resulting `ElementTree` object has a `findall()` method that will iterate through the available values.

We'll be looking for constructs, such as the following XML example:

```
<Placemark><Point>
<coordinates>-76.33029518659048,
    37.54901619777347,0</coordinates>
</Point></Placemark>
```

The file will have a number of `<Placemark>` tags, each of which has a point and coordinate structure within it. This is typical of **Keyhole Markup Language** (**KML**) files that contain geographic information.

Parsing an XML file can be approached at two levels of abstraction. At the lower level, we need to locate the various tags, attribute values, and content within the XML file. At a higher level, we want to make useful objects out of the text and attribute values.

The lower-level processing can be approached in the following way:

```
import xml.etree.ElementTree as XML
from typing import Text, List, TextIO, Iterable
def row_iter_kml(file_obj: TextIO) -> Iterable[List[Text]]:
    ns_map= {
        "ns0": "http://www.opengis.net/kml/2.2",
        "ns1": "http://www.google.com/kml/ext/2.2"}
    path_to_points= ("./ns0:Document/ns0:Folder/ns0:Placemark/"
                "ns0:Point/ns0:coordinates")
    doc= XML.parse(file_obj)
    return (comma_split(Text(coordinates.text))
            for coordinates in
            doc.findall(path_to_points, ns_map))
```

This function requires text from a file opened via a `with` statement. The result is a generator that creates list objects from the latitude/longitude pairs. As a part of the XML processing, this function includes a simple static `dict` object, `ns_map`, that provides the `namespace` mapping information for the XML tags we'll be searching. This dictionary will be used by the `ElementTree.findall()` method.

The essence of the parsing is a generator function that uses the sequence of tags located by `doc.findall()`. This sequence of tags is then processed by a `comma_split()` function to tease the text value into its comma-separated components.

The `comma_split()` function is the functional version of the `split()` method of a string, which is as follows:

```
def comma_split(text: Text) -> List[Text]:
    return text.split(",")
```

We've used the functional wrapper to emphasize a slightly more uniform syntax. We've also added explicit type hints to make it clear that text is converted to a list of text values. Without the type hint, there are two potential definitions of `split()` that could be meant. The method applies to `bytes` as well as `str`. We've used the `Text` type name, which is an alias for `str` in Python 3.

The result of the `row_iter_kml()` function is an iterable sequence of rows of data. Each row will be a list composed of three strings—`latitude`, `longitude`, and `altitude` of a way point along this path. This isn't directly useful yet. We'll need to do some more processing to get `latitude` and `longitude` as well as converting these two numbers into useful floating-point values.

This idea of an iterable sequence of tuples (or lists) allows us to process some kinds of data files in a simple and uniform way. In `Chapter 3`, *Functions, Iterators, and Generators,* we looked at how **Comma Separated Values** (**CSV**) files are easily handled as rows of tuples. In `Chapter 6`, *Recursions and Reductions,* we'll revisit the parsing idea to compare these various examples.

The output from the preceding function looks like the following example:

```
[['-76.33029518659048', '37.54901619777347', '0'],
 ['-76.27383399999999', '37.840832', '0'],
 ['-76.459503', '38.331501', '0'],
 etc.
 ['-76.47350299999999', '38.976334', '0']]
```

Each row is the source text of the `<ns0:coordinates>` tag split using the (`,`) that's part of the text content. The values are the east-west longitude, north-south latitude, and altitude. We'll apply some additional functions to the output of this function to create a usable subset of this data.

Parsing a file at a higher level

Once we've parsed the low-level syntax to transform XML to Python, we can restructure the raw data into something usable in our Python program. This kind of structuring applies to XML, **JavaScript Object Notation** (**JSON**), CSV, and any of the wide variety of physical formats in which data is serialized.

We'll aim to write a small suite of generator functions that transforms the parsed data into a form our application can use. The generator functions include some simple transformations on the text that are found by the `row_iter_kml()` function, which are as follows:

- Discarding `altitude` can also be stated as keeping only `latitude` and `longitude`
- Changing the order from (`longitude, latitude`) to (`latitude, longitude`)

We can make these two transformations have more syntactic uniformity by defining a utility function, as follows:

```
def pick_lat_lon(
  lon: Text, lat: Text, alt: Text) -> Tuple[Text, Text]:
    return lat, lon
```

We've created a function to take three argument values and created a tuple from two of them. The type hints are more complex than the function itself.

We can use this function as follows:

```
from typing import Text, List, Iterable

Rows = Iterable[List[Text]]
LL_Text = Tuple[Text, Text]
def lat_lon_kml(row_iter: Rows) -> Iterable[LL_Text]:
    return (pick_lat_lon(*row) for row in row_iter)
```

This function will apply the `pick_lat_lon()` function to each row from a source iterator. We've used `*row` to assign each element of the row-three tuple to separate parameters of the `pick_lat_lon()` function. The function can then extract and reorder the two relevant values from each three-tuple.

To simplify the function definition, we've defined two type aliases: `Rows` and `LL_Text`. These type aliases can simplify a function definition. They can also be reused to ensure that several related functions are all working with the same types of objects.

This kind of functional design allows us to freely replace any function with its equivalent, which makes refactoring quite simple. We tried to achieve this goal when we provided alternative implementations of the various functions. In principle, a clever functional language compiler may do some replacements as a part of an optimization pass.

These functions can be combined to parse the file and build a structure we can use. Here's an example of some code that could be used for this purpose:

```
url = "file:./Winter%202012-2013.kml"
with urllib.request.urlopen(url) as source:
    v1= tuple(lat_lon_kml(row_iter_kml(source)))
print(v1)
```

This script uses the `urllib` command to open a source. In this case, it's a local file. However, we can also open a KML file on a remote server. Our objective in using this kind of file opening is to ensure that our processing is uniform no matter what the source of the data is.

The script is built around the two functions that do low-level parsing of the KML source. The `row_iter_kml(source)` expression produces a sequence of text columns. The `lat_lon_kml()` function will extract and reorder the `latitude` and `longitude` values. This creates an intermediate result that sets the stage for further processing. The subsequent processing is independent of the original format.

When we run this, we see results such as the following:

```
(('37.54901619777347', '-76.33029518659048'),
 ('37.840832', '-76.27383399999999'),
 ('38.331501', '-76.459503'),
 ('38.330166', '-76.458504'),
 ('38.976334', '-76.47350299999999'))
```

We've extracted just the `latitude` and `longitude` values from a complex XML file using an almost purely functional approach. As the result is iterable, we can continue to use functional programming techniques to process each point that we retrieve from the file.

We've explicitly separated low-level XML parsing from higher-level reorganization of the data. The XML parsing produced a generic tuple of string structure. This is compatible with the output from the CSV parser. When working with SQL databases, we'll have a similar iterable of tuple structures. This allows us to write code for higher-level processing that can work with data from a variety of sources.

We'll show you a series of transformations to re-arrange this data from a collection of strings to a collection of waypoints along a route. This will involve a number of transformations. We'll need to restructure the data as well as convert from `strings` to `floating-point` values. We'll also look at a few ways to simplify and clarify the subsequent processing steps. We'll use this dataset in later chapters because it's quite complex.

Pairing up items from a sequence

A common restructuring requirement is to make start-stop pairs out of points in a sequence. Given a sequence, $S = \{s_0, s_1, s_2, \ldots, s_n\}$, we would also want to create a paired sequence, $\hat{S} = \{(s_0, s_1), (s_1, s_2), (s_2, s_3), \ldots, (s_{n-1}, s_n)\}$. The first and second items form a pair. The second and third items form the next pair. When doing time-series analysis, we may be combining more widely separated values. In this example, the pairs are immediately adjacent values.

A paired sequence will allow us to use each pair to compute distances from point to point using a trivial application of a `haversine` function. This technique is also used to convert a path of points into a series of line segments in a graphics, application.

Why pair up items? Why not do something such as this:

```
begin= next(iterable)
for end in iterable:
    compute_something(begin, end)
    begin = end
```

This, clearly, will process each leg of the data as a begin-end pair. However, the processing function and the loop that restructures the data are tightly bound, making reuse more complex than necessary. The algorithm for pairing is hard to test in isolation because it's bound to the `compute_something()` function.

This combined function also limits our ability to reconfigure the application. There's no easy way to inject an alternative implementation of the `compute_something()` function. Additionally, we've got a piece of an explicit state, the `begin` variable, which makes life potentially complex. If we try to add features to the body of `loop`, we can easily fail to set the `begin` variable correctly if a point is dropped from consideration. A `filter()` function introduces an `if` statement that can lead to an error in updating the `begin` variable.

We achieve better reuse by separating this simple pairing function. This, in the long run, is one of our goals. If we build up a library of helpful primitives such as this pairing function, we can tackle problems more quickly and confidently.

There are many ways to pair up the points along the route to create start and stop information for each leg. We'll look at a few here and then revisit this in `Chapter 5`, *Higher-Order Functions*, and again in `Chapter 7`, *The Itertools Module*. Creating pairs can be done in a purely functional way using a recursion.

The following code is one version of a function to pair up the points along a route:

```
from typing import Iterator, Any
Item_Iter = Iterator[Any]
Pairs_Iter = Iterator[Tuple[float, float]]
def pairs(iterator: Item_Iter) -> Pairs_Iter:
    def pair_from(
            head: Any,
            iterable_tail: Item_Iter) -> Pairs_Iter:
        nxt= next(iterable_tail)
        yield head, nxt
        yield from pair_from(nxt, iterable_tail)

    try:
        return pair_from(next(iterator), iterator)
    except StopIteration:
         return iter([])
```

The essential work is done by the internal `pair_from()` function. This works with the item at the head of an iterator plus the iterator object itself. It yields the first pair, pops the next item from the iterable, and then invokes itself recursively to yield any additional pairs.

The type hints state the parameter, `iterator`, must be of type `Item_Iter`. The result is of the `Pairs_Iter` type, an iterator over two-tuples, where each item is a `float` type. These are hints used by the **mypy** program to check that our code is likely to work. The type hint declarations are contained in the `typing` module.

The input must be an iterator that responds to the `next()` function. To work with a collection object, the `iter()` function must be used explicitly to create an iterator from the collection.

We've invoked the `pair_from()` function from the `pairs()` function. The `pairs()` function ensures that the initialization is handled properly by getting the initial item from the iterator argument. In the rare case of an empty iterator, the initial call to `next()` will raise a `StopIteration` exception; this situation will create an empty iterable.

Python's iterable recursion involves a `for` loop to properly consume and yield the results from the recursion. If we try to use a simpler-looking `return pair_from(nxt, iterable_tail)` statement, we'll see that it does not properly consume the iterable and yield all of the values. Recursion in a generator function requires `yield` from a statement to consume the resulting iterable. For this, use `yield from recursive_iter(args)`. Something like `return recursive_iter(args)` will return only a generator object; it doesn't evaluate the function to return the generated values.

Our strategy for performing tail-call optimization is to replace the recursion with a generator expression. We can clearly optimize this recursion into a simple `for` loop. The following code is another version of a function to pair up the points along a route:

```
from typing import Iterator, Any, Iterable, TypeVar
T_ = TypeVar('T_')
Pairs_Iter = Iterator[Tuple[T_, T_]]
def legs(lat_lon_iter: Iterator[T_]) -> Pairs_Iter:
    begin = next(lat_lon_iter)
    for end in lat_lon_iter:
        yield begin, end
        begin = end
```

The version is quite fast and free from stack limits. It's independent of any particular type of sequence, as it will pair up anything emitted by a sequence generator. As there's no processing function inside the loop, we can reuse the `legs()` function as needed.

The type variable, `T_`, created with the `TypeVar` function, is used to clarify precisely how the `legs()` function restructures the data. The hint says that the input type is preserved on output. The input type is an `Iterator` of some arbitrary type, `T_`; the output will include tuples of the same type, `T_`. No other conversion is implied by the function.

The `begin` and `end` variables maintain the state of the computation. The use of stateful variables doesn't fit the ideal of using immutable objects for functional programming. The optimization is important. It's also invisible to users of the function, making it a Pythonic functional hybrid.

We can think of this function as one that yields the following kind of sequence of pairs:

```
list[0:1], list[1:2], list[2:3], ..., list[-2:]
```

Another view of this function is as follows:

```
zip(list, list[1:])
```

While informative, this zip-based version only work for sequence objects. The `legs()` and `pairs()` functions work for any iterable, including sequence objects.

Using the iter() function explicitly

The purely functional viewpoint is that all of our iterables can be processed with recursive functions, where the state is merely the recursive call stack. Pragmatically, Python iterables will often involve evaluation of other `for` loops. There are two common situations: collection objects and iterables. When working with a collection object, an iterator object is created by the `for` statement. When working with a generator function, the generator function is an iterator and maintains its own internal state. Often, these are equivalent, from a Python programming perspective. In rare cases—generally those situations where we have to use an explicit `next()` function—two won't be precisely equivalent.

The `legs()` function shown previously has an explicit `next()` evaluation to get the first value from the iterable. This works wonderfully well with generator functions, expressions, and other iterables. It doesn't work with sequence objects such as tuples or `lists`.

The following code contains three examples to clarify the use of the `next()` and `iter()` functions:

```
>>> list(legs(x for x in range(3)))
[(0, 1), (1, 2)]
>>> list(legs([0,1,2]))
Traceback (most recent call last):
  File "<stdin>", line 1, in <module>
  File "<stdin>", line 2, in legs
TypeError: 'list' object is not an iterator
>>> list(legs(iter([0,1,2])))
[(0, 1), (1, 2)]
```

In the first case, we applied the `legs()` function to an iterable. In this case, the iterable was a generator expression. This is the expected behavior based on our previous examples in this chapter. The items are properly paired up to create two legs from three waypoints.

In the second case, we tried to apply the `legs()` function to a sequence. This resulted in an error. While a `list` object and an iterable are equivalent when used in a `for` statement, they aren't equivalent everywhere. A sequence isn't an iterator; it doesn't implement the `next()` function. The `for` statement handles this gracefully, however, by creating an iterator from a sequence automatically.

To make the second case work, we need to explicitly create an iterator from a `list` object. This permits the `legs()` function to get the first item from the iterator over the `list` items. The `iter()` function will create an iterator from a list.

Extending a simple loop

We have two kinds of extensions we could factor into a simple loop. We'll look first at a `filter` extension. In this case, we may be rejecting values from further consideration. They may be data outliers, or perhaps source data that's improperly formatted. Then, we'll look at mapping source data by performing a simple transformation to create new objects from the original objects. In our case, we'll be transforming strings to floating-point numbers. The idea of extending a simple `for` statement with a mapping, however, applies to many situations. We'll look at refactoring the above `pairs()` function. What if we need to adjust the sequence of points to discard a value? This will introduce a `filter` extension that rejects some data values.

The loop we're designing simply returns pairs without performing any additional application-related processing—the complexity is minimal. Simplicity means we're somewhat less likely to confuse the processing state.

Adding a `filter` extension to this design could look something like the following code snippet:

```
from typing import Iterator, Any, Iterable
Pairs_Iter = Iterator[Tuple[float, float]]
LL_Iter = Iterable[
    Tuple[Tuple[float, float], Tuple[float, float]]]
def legs_filter(lat_lon_iter: Pairs_Iter) -> LL_Iter:
    begin = next(lat_lon_iter)
    for end in lat_lon_iter:
        if #some rule for rejecting:
            continue
        yield begin, end
        begin = end
```

We have plugged in a processing rule to reject certain values. As the `loop` remains succinct and expressive, we are confident that the processing will be done properly. Also, we can easily write a test for this function, as the results work for any iterable, irrespective of the long-term destination of the pairs.

We haven't really provided much information about the `#some rule for rejecting` code. This is a kind of condition that uses `begin`, `end`, or both variables to reject the point from further consideration. For example, it may reject `begin == end` to avoid zero-length legs.

The next refactoring will introduce additional mapping to a loop. Adding mappings is common when a design is evolving. In our case, we have a sequence of `string` values. We need to convert these to `float` values for later use. This is a relatively simple mapping that shows the design pattern.

The following is one way to handle this data mapping, through a generator expression that wraps a generator function:

```
trip = list(
    legs(
        (float(lat), float(lon))
        for lat,lon in lat_lon_kml(row_iter_kml(source))
    )
)
```

We've applied the `legs()` function to a generator expression that creates `float` values from the output of the `lat_lon_kml()` function. We can read this in the opposite order as well. The `lat_lon_kml()` function's output is transformed into a pair of `float` values, which is then transformed into a sequence of `legs`.

This is starting to get complex. We've got a large number of nested functions here. We're applying `float()`, `legs()`, and `list()` to a data generator. One common way of refactoring complex expressions is to separate the generator expression from any materialized collection. We can do the following to simplify the expression:

```
ll_iter = (
    (float(lat), float(lon))
    for lat,lon in lat_lon_kml(row_iter_kml(source))
)
print(tuple(legs(ll_iter)))
```

We've assigned the generator function to a variable named `ll_iter`. This variable isn't a collection object; it's a generator of item. We're not using a `list` comprehension to create an object. We've merely assigned the generator expression to a variable name. We've then used the `flt` variable in another expression.

The evaluation of the `tuple()` method actually leads to a proper object being built so that we can print the output. The `flt` variable's objects are created only as needed.

There is other refactoring we might like to do. In general, the source of the data is something we often want to change. In our example, the `lat_lon_kml()` function is tightly bound in the rest of the expression. This makes reuse difficult when we have a different data source.

In the case where the `float()` operation is something we'd like to parameterize so that we can reuse it, we can define a function around the generator expression. We'll extract some of the processing into a separate function merely to group the operations. In our case, the string-pair to float-pair is unique to particular source data. We can rewrite a complex float-from-string expression into a simpler function, such as:

```
from typing import Iterator, Tuple, Text, Iterable
Text_Iter = Iterable[Tuple[Text, Text]]
LL_Iter = Iterable[Tuple[float, float]]
def float_from_pair(lat_lon_iter: Text_Iter) -> LL_Iter:
    return (
        (float(lat), float(lon))
        for lat,lon in lat_lon_iter
    )
```

The `float_from_pair()` function applies the `float()` function to the first and second values of each item in the iterable, yielding a two-tuple of floats created from an input value. We've relied on Python's `for` statement to decompose the two-tuple.

The type hints insist that the input matches the `Text_Iter` type alias—it must be an iterable source of pairs of `Text` values. The result uses the `LL_Iter` type alias—this must be an iterable of pairs of `float` values. The `LL_Iter` type alias may be used elsewhere in a complex set of function definitions.

We can use this function in the following context:

```
legs(
    float_from_pair(
        lat_lon_kml(
            row_iter_kml(source))))
```

We're going to create `legs` that are built from `float` values that come from a KML file. It's fairly easy to visualize the processing, as each stage in the process is a simple prefix function. Each function's input is the output from the next function in the nested processing steps.

When parsing, we often have sequences of `string` values. For numeric applications, we'll need to convert `strings` to `float`, `int`, or `Decimal` values. This often involves inserting a function such as the `float_from_pair()` function into a sequence of expressions that clean up the source data.

Our previous output was all strings; it looked like the following code snippet:

```
(('37.54901619777347', '-76.33029518659048'),
 ('37.840832', '-76.27383399999999'),
 ...
 ('38.976334', '-76.47350299999999'))
```

We'll want data like the following code snippet, where we have floats:

```
(((37.54901619777347, -76.33029518659048),
 (37.840832, -76.273834)), ((37.840832, -76.273834),
 ...
 ((38.330166, -76.458504), (38.976334, -76.473503)))
```

We'll need to create a pipeline of simpler transformation functions. Here, we arrived at `flt= ((float(lat), float(lon)) for lat,lon in lat_lon_kml(...))`. We can exploit the substitution rule for functions and replace a complex expression such as `(float(lat), float(lon)) for lat,lon in lat_lon_kml(...))` with a function that has the same value, in this case `float_from_pair(lat_lon_kml(...))`. This kind of refactoring allows us to be sure that the simplification has the same effect as a more complex expression.

There are some simplifications that we'll look at in `Chapter 5`, *Higher-Order Functions*. We will revisit this in `Chapter 6`, *Recursions and Reductions*, to see how to apply these simplifications to the file-parsing problem.

Applying generator expressions to scalar functions

We'll look at a more complex kind of generator expression to map data values from one kind of data to another. In this case, we'll apply a fairly complex function to individual data values created by a generator.

We'll call these non-generator functions **scalar**, as they work with simple atomic values. To work with collections of data, a scalar function will be embedded in a generator expression.

To continue the example started earlier, we'll provide a `haversine` function and then use a generator expression to apply a scalar `haversine()` function to a sequence of pairs from our KML file.

The `haversine()` function looks like the following code:

```
from math import radians, sin, cos, sqrt, asin
from typing import Tuple
MI= 3959
NM= 3440
KM= 6371
Point = Tuple[float, float]
def haversine(p1: Point, p2: Point, R: float=NM) -> float:
    lat_1, lon_1= p1
    lat_2, lon_2= p2
    Δ_lat = radians(lat_2 - lat_1)
    Δ_lon = radians(lon_2 - lon_1)
    lat_1 = radians(lat_1)
    lat_2 = radians(lat_2)
    a = sqrt(
        sin(Δ_lat/2)**2 +
        cos(lat_1)*cos(lat_2)*sin(Δ_lon/2)**2
    )
    c = 2*asin(a)
    return R * c
```

This is a relatively simple implementation copied from the World Wide Web. The start and end points have type hints. The return value is also provided with a hint. The explicit use of `Point = Tuple[float, float]` makes it possible for the **mypy** tool to confirm that this function is used properly.

The following code is how we could use our collection of functions to examine some KML data and produce a sequence of distances:

```
trip= (
    (start, end, round(haversine(start, end),4))
    for start,end in
        legs(float_from_pair(lat_lon_kml()))
)
for start, end, dist in trip:
    print(start, end, dist)
```

The essence of the processing is the generator expression assigned to the `trip` variable. We've assembled three tuples with a start, end, and the distance from start to end. The start and end pairs come from the `legs()` function. The `legs()` function works with `floating-point` data built from the `latitude-longitude` pairs extracted from a KML file.

The output looks like the following command snippet:

```
(37.54901619777347, -76.33029518659048) (37.840832, -76.273834) 17.7246
(37.840832, -76.273834) (38.331501, -76.459503) 30.7382
(38.331501, -76.459503) (38.845501, -76.537331) 31.0756
(36.843334, -76.298668) (37.549, -76.331169) 42.3962
(37.549, -76.331169) (38.330166, -76.458504) 47.2866
(38.330166, -76.458504) (38.976334, -76.473503) 38.8019
```

Each individual processing step has been defined succinctly. The overview, similarly, can be expressed succinctly as a composition of functions and generator expressions.

Clearly, there are several further processing steps we may like to apply to this data. The first, of course, is to use the `format()` method of a string to produce better-looking output.

More importantly, there are a number of aggregate values we'd like to extract from this data. We'll call these values reductions of the available data. We'd like to reduce the data to get the maximum and minimum latitude, for example, to show the extreme north and south ends of this route. We'd like to reduce the data to get the maximum distance in one leg as well as the total distance for all `legs`.

The problem we'll have using Python is that the output generator in the `trip` variable can be used only once. We can't easily perform several reductions of this detailed data. We can use `itertools.tee()` to work with the iterable several times. It seems wasteful, however, to read and parse the KML file for each reduction.

We can make our processing more efficient by materializing intermediate results. We'll look at this in the next section. Then we will see how to compute multiple reductions of the available data.

Using any() and all() as reductions

The `any()` and `all()` functions provide `boolean` reduction capabilities. Both functions reduce a collection of values to a single `True` or `False`. The `all()` function ensures that all values are `True`. The `any()` function ensures that at least one value is `True`.

These functions are closely related to a universal quantifier and an existential quantifier used to express mathematical logic. We may, for example, want to assert that all elements in a given collection have a property. One formalism for this could look like the following:

$$(\forall_{x \in S})\text{Prime}(x)$$

We read this as *for all x in S, the function, Prime(x), is true*. We've put a quantifier, for all, in front of the logical expression.

In Python we switch the order of the items slightly to transcribe the logic expression as follows:

```
all(isprime(x) for x in someset)
```

This will evaluate the `isprime(x)` function for each distinct value of `x` and reduce the collection of values to a single `True` or `False`.

The `any()` function is related to the existential quantifier. If we want to assert that no value in a collection is prime, we could use one of these two equivalent expressions:

$$\neg(\forall_{x \in S})\text{Prime}(x) \equiv (\exists_{x \in S})\neg\text{Prime}(x)$$

The first states that it is not the case that all elements in *S* are prime. The second version asserts that there exists one element in *S* that is not prime. These two are equivalent, that is if not all elements are prime, then one element must be non-prime.

In Python, we can switch the order of the terms and transcribe these to working code as follows:

```
not all(isprime(x) for x in someset)
any(not isprime(x) for x in someset)
```

As they're equivalent, there are two reasons for preferring one over the other: performance and clarity. The performance is nearly identical, so it boils down to clarity. Which of these states the condition the most clearly?

The `all()` function can be described as an `and` reduction of a set of values. The result is similar to folding the `and` operator between the given sequence of values. The `any()` function, similarly, can be described as an `or` reduction. We'll return to this kind of general purpose reducing when we look at the `reduce()` function in `Chapter 10`, *The Functools Module*. There's no best answer here; it's a question of what seems most readable to the intended audience.

We also need to look at the degenerate case of these functions. What if the sequence has no elements? What are the values of `all(())` or `all([])`?

If we ask, "Are all the elements in an empty set prime? Then what's the answer? For guidance on this, we'll expand the question slightly, and look at the idea of an identity element.

If we ask, "Are all the elements in an empty set prime, and are all the elements in `SomeSet` prime?" we have a hint as to how we have to proceed. We're performing an `and` reduction of an empty set and an `and` reduction of `SomeSet`:

$(\forall_{x \in \emptyset})\text{Prime}(x) \wedge (\forall_{x \in S})\text{Prime}(x)$

It turns out that the `and` operator can be distributed freely. We can rewrite this to, as a union of the two sets, which is then evaluated for being prime:

$(\forall_{x \in \emptyset \cup S})\text{Prime}(x)$

Clearly, $S \cup \emptyset \equiv S$. If we union a set, *S*, with an empty set, we get the original set, *S*. The empty set can be called the **union identify element**. This is parallel to the way zero is the additive identity element:

$$a + 0 = a$$

Similarly, `any(())` must be the `or` identity element, which is `False`. If we think of the multiplicative identify element, 1, where $b \times 1 = b$, then `all(())` must be `True`.

The following code demonstrates that Python follows these rules:

```
>>> all(())
True
>>> any(())
False
```

Python gives us some very nice tools to perform processing that involves logic. We have the built-in `and`, `or`, and `not` operators. However, we also have these collection-oriented `any()` and `all()` functions.

Using len() and sum()

The `len()` and `sum()` functions provide two simple reductions—a count of the elements and the sum of the elements in a sequence. These two functions are mathematically similar, but their Python implementation is quite different.

Mathematically, we can observe this cool parallelism. The `len()` function returns the sum of ones for each value in a collection, $X: \sum_{x \in X} 1 = \sum_{x \in X} x^0$.

The `sum()` function returns the sum of *x* for each value in a collection, $X: \sum_{x \in X} x = \sum_{x \in X} x^1$.

The `sum()` function works for any iterable. The `len()` function doesn't apply to iterables; it only applies to sequences. This little asymmetry in the implementation of these functions is a little awkward around the edges of statistical algorithms.

For empty sequences, both of these functions return a proper additive identity element of zero:

```
>>> sum(())
0
```

Of course, `sum(())` returns an integer zero. When other numeric types are used, the integer zero will be coerced to the proper type for the available data.

Using sums and counts for statistics

The definitions of the arithmetic mean have an appealingly trivial definition based on `sum()` and `len()`, which is as follows:

```
def mean(items):
    return sum(items)/len(items)
```

While elegant, this doesn't actually work for iterables. This definition only works for collections that support the `len()` function. This is easy to discover when trying to write proper type annotations. The definition of `mean(items: Iterable)->float` won't work because `Iterable` types don't support `len()`.

Indeed, we have a hard time performing a simple computation of mean or standard deviation based on iterables. In Python, we must either materialize a sequence object or resort to somewhat more complex operations.

The definition needs to look like this:

```
from collections import Sequence
def mean(items: Sequence) -> float:
    return sum(items)/len(items)
```

This includes the appropriate type hints to assure that `sum()` and `len()` will both work.

We have some alternative and elegant expressions for mean and standard deviation in the following definitions:

```
import math
s0 = len(data)   # sum(x**0 for x in data)
s1 = sum(data)   # sum(x**1 for x in data)
s2 = sum(x**2 for x in data)
mean = s1/s0
stdev = math.sqrt(s2/s0 - (s1/s0)**2)
```

These three sums, `s0`, `s1`, and `s2`, have a tidy, parallel structure. We can easily compute the mean from two of the sums. The standard deviation is a bit more complex, but it's based on the three available sums.

This kind of pleasant symmetry also works for more complex statistical functions, such as correlation and even least-squares linear regression.

The moment of correlation between two sets of samples can be computed from their standardized value. The following is a function to compute the standardized value:

```
def z(x: float, m_x: float, s_x: float) -> float:
    return (x-m_x)/s_x
```

The calculation is simply to subtract the mean, μ_x, from each sample, `x`, and divide by the standard deviation, σ_x. This gives us a value measured in units of sigma, σ. A value **±1 σ** is expected about two-thirds of the time. Larger values should be less common. A value outside **±3 σ** should happen less than one percent of the time.

We can use this scalar function as follows:

```
>>> d = [2, 4, 4, 4, 5, 5, 7, 9]
>>> list(z(x, mean(d), stdev(d)) for x in d)
[-1.5, -0.5, -0.5, -0.5, 0.0, 0.0, 1.0, 2.0]
```

We've materialized a `list` that consists of normalized scores based on some raw data in the variable, `d`. We used a generator expression to apply the scalar function, `z()`, to the sequence object.

The `mean()` and `stdev()` functions are simply based on the examples shown previously:

```
def mean(samples: Sequence) -> float:
    return s1(samples)/s0(samples)
def stdev(samples: Sequence) -> float:
    N= s0(samples)
    return sqrt((s2(samples)/N)-(s1(samples)/N)**2)
```

The three sum functions, similarly, can be defined, instead of the lambda forms, as shown in the following code:

```
def s0(samples: Sequence) -> float:
    return sum(1 for x in samples) # or len(data)
def s1(samples: Sequence) -> float:
    return sum(x for x in samples) # or sum(data)
def s2(samples: Sequence) -> float:
    return sum(x*x for x in samples)
```

While this is very expressive and succinct, it's a little frustrating because we can't simply use an iterable here. When computing a mean, both a sum of the iterable and a count of the iterable are required. For the standard deviation, two sums and a count of the iterable are all required. For this kind of statistical processing, we must materialize a sequence object so that we can examine the data multiple times.

The following code shows how we can compute the correlation between two sets of samples:

```
def corr(samples1: Sequence, samples2: Sequence) -> float:
    m_1, s_1 = mean(samples1), stdev(samples1)
    m_2, s_2 = mean(samples2), stdev(samples2)
    z_1 = (z( x, m_1, s_1 ) for x in samples1)
    z_2 = (z( x, m_2, s_2 ) for x in samples2)
    r = (sum(zx1*zx2 for zx1, zx2 in zip(z_1, z_2))
        / len(samples1))
    return r
```

This correlation function gathers basic statistical summaries of the two sets of samples: the mean and standard deviation. Given these summaries, we define two generator functions that will create normalized values for each set of samples. We can then use the `zip()` function (see the next example) to pair up items from the two sequences of normalized values and compute the product of those two normalized values. The average of the product of the normalized scores is the correlation.

The following code is an example of gathering the correlation between two sets of samples:

```
>>> # Height (m)
>>> xi= [1.47, 1.50, 1.52, 1.55, 1.57, 1.60, 1.63, 1.65,
...     1.68, 1.70, 1.73, 1.75, 1.78, 1.80, 1.83,]
>>> # Mass (kg)
>>> yi= [52.21,53.12,54.48,55.84,57.20,58.57,59.93,61.29,
...     63.11, 64.47, 66.28, 68.10, 69.92, 72.19, 74.46,]
>>> round(corr( xi, yi ), 5)
0.99458
```

We've shown you two sequences of data points, `xi` and `yi`. The correlation is over **.99**, which shows a very strong relationship between the two sequences.

This shows us one of the strengths of functional programming. We've created a handy statistical module using a half-dozen functions with definitions that are single expressions. The counterexample is the `corr()` function that can be reduced to a single very long expression. Each internal variable in this function is used just once; a local variable can be replaced with a copy and paste of the expression that created it. This shows us that the `corr()` function has a functional design, even though it's written out in six separate lines of Python.

Using zip() to structure and flatten sequences

The `zip()` function interleaves values from several iterators or sequences. It will create *n* tuples from the values in each of the *n* input iterables or sequences. We used it in the previous section to interleave data points from two sets of samples, creating two-tuples.

The `zip()` function is a generator. It does not materialize a resulting collection.

The following is an example of code that shows what the `zip()` function does:

```
>>> xi= [1.47, 1.50, 1.52, 1.55, 1.57, 1.60, 1.63, 1.65,
... 1.68, 1.70, 1.73, 1.75, 1.78, 1.80, 1.83,]
>>> yi= [52.21, 53.12, 54.48, 55.84, 57.20, 58.57, 59.93, 61.29,
... 63.11, 64.47, 66.28, 68.10, 69.92, 72.19, 74.46,]
>>> zip( xi, yi )
<zip object at 0x101d62ab8>
```

```
>>> list(zip( xi, yi ))
[(1.47, 52.21), (1.5, 53.12), (1.52, 54.48),
 (1.55, 55.84), (1.57, 57.2), (1.6, 58.57),
 (1.63, 59.93), (1.65, 61.29), (1.68, 63.11),
 (1.7, 64.47), (1.73, 66.28), (1.75, 68.1),
 (1.78, 69.92), (1.8, 72.19), (1.83, 74.46)]
```

There are a number of edge cases for the `zip()` function. We must ask the following questions about its behavior:

- What happens where then are no arguments at all?
- What happens where there's only one argument?
- What happens when the sequences are different lengths?

As with other functions, such as `any()`, `all()`, `len()`, and `sum()`, we want an identity value as a result when applying the reduction to an empty sequence. For example, `sum(())` should be zero. This concept tells us what the identity value for `zip()` should be.

Clearly, each of these edge cases must produce some kind of iterable output. Here are some examples of code that clarify the behaviors. First, the empty argument list:

```
>>> zip()
<zip object at 0x101d62ab8>
>>> list(_)
[]
```

We can see that the `zip()` function with no arguments is a generator function, but there won't be any items. This fits the requirement that the output is iterable.

Next, we'll try a single iterable:

```
>>> zip( (1,2,3) )
<zip object at 0x101d62ab8>
>>> list(_)
[(1,), (2,), (3,)]
```

In this case, the `zip()` function emitted one tuple from each input value. This too makes considerable sense.

Finally, we'll look at the different-length `list` approach used by the `zip()` function:

```
>>> list(zip((1, 2, 3), ('a', 'b')))
[(1, 'a'), (2, 'b')]
```

This result is debatable. Why truncate? Why not pad the shorter list with `None` values? This alternate definition of the `zip()` function is available in the `itertools` module as the `zip_longest()` function. We'll look at this in `Chapter 8`, *The Itertools Module*.

Unzipping a zipped sequence

We can insert `zip()` mapping can be inverted. We'll look at several ways to unzip a collection of tuples.

We can't fully unzip an iterable of tuples, since we might want to make multiple passes over the data. Depending on our needs, we may need to materialize the iterable to extract multiple values.

The first way is something we've seen many times: we can use a generator function to unzip a sequence of tuples. For example, assume that the following pairs are a sequence object with two-tuples:

```
p0= (x[0] for x in pairs)
p1= (x[1] for x in pairs)
```

This will create two sequences. The `p0` sequence has the first element of each two-tuple; the `p1` sequence has the second element of each two-tuple.

Under some circumstances, we can use the multiple assignment of a `for` loop to decompose the tuples. The following is an example that computes the sum of the products:

```
sum(p0*p1 for for p0, p1 in pairs)
```

We used the `for` statement to decompose each two-tuple into `p0` and `p1`.

Flattening sequences

Sometimes, we'll have zipped data that needs to be flattened. For example, our input could be a file that has columnar data. It looks like this:

```
2    3    5    7    11   13   17   19   23   29
31   37   41   43   47   53   59   61   67   71
...
```

We can easily use `(line.split() for line in file)` to create a sequence. Each item within that sequence will be a 10-item tuple from the values on a single line.

This creates data in blocks of 10 values. It looks as follows:

```
>>> blocked = list(line.split() for line in file)
>>> blocked
[['2', '3', '5', '7', '11', '13', '17', '19', '23', '29'], ['31', '37',
'41', '43', '47', '53', '59', '61', '67', '71'], ['179', '181', '191',
'193', '197', '199', '211', '223', '227', '229']]
```

This is a start, but it isn't complete. We want to get the numbers into a single, flat sequence. Each item in the input is a 10 tuple; we'd rather not deal with decomposing this one item at a time.

We can use a two-level generator expression, as shown in the following code snippet, for this kind of flattening:

```
>>> (x for line in blocked for x in line)
<generator object <genexpr> at 0x101cead70>
>>> list(_)
['2', '3', '5', '7', '11', '13', '17', '19', '23', '29', '31',
 '37', '41', '43', '47', '53', '59', '61', '67', '71',
 ... ]
```

The first `for` clause assigns each item—a list of 10 values— from the `blocked` list to the `line` variable. The second `for` clause assigns each individual string from the `line` variable to the `x` variable. The final generator is this sequence of values assigned to the `x` variable.

We can understand this via a simple rewrite as follows:

```
def flatten(data: Iterable[Iterable[Any]]) -> Iterable[Any]:
    for line in data:
        for x in line:
            yield x
```

This transformation shows us how the generator expression works. The first `for` clause (`for line in data`) steps through each 10-tuple in the data. The second `for` clause (`for x in line`) steps through each item in the first `for` clause.

This expression flattens a sequence-of-sequence structure into a single sequence. More generally, it flattens any iterable that contains an iterable into a single, flat iterable. It will work for list-of-list as well as list-of-set or any other combination of nested iterables.

Structuring flat sequences

Sometimes, we'll have raw data that is a flat list of values that we'd like to bunch up into subgroups. This is a bit more complex. We can use the `itertools` module's `groupby()` function to implement this. This will have to wait until `Chapter 8`, *The Iterools Module*.

Let's say we have a simple flat `list`, as follows:

```
flat= ['2', '3', '5', '7', '11', '13', '17', '19', '23', '29',
 '31', '37', '41', '43', '47', '53', '59', '61', '67', '71',
 ... ]
```

We can write nested generator functions to build a sequence-of-sequence structure from flat data. To do this, we'll need a single iterator that we can use multiple times. The expression looks like the following code snippet:

```
>>> flat_iter = iter(flat)
>>> (tuple(next(flat_iter) for i in range(5))
...      for row in range(len(flat)//5)
... )
<generator object <genexpr> at 0x101cead70>
>>> list(_)
[('2', '3', '5', '7', '11'),
 ('13', '17', '19', '23', '29'),
 ('31', '37', '41', '43', '47'),
 ('53', '59', '61', '67', '71'),
 ...
]
```

First, we create an iterator that exists outside either of the two loops that we'll use to create our sequence-of-sequences. The generator expression uses `tuple(next(flat_iter) for i in range(5))` to create five-item tuples from the iterable values in the `flat_iter` variable. This expression is nested inside another generator that repeats the inner loop the proper number of times to create the required sequence of values.

This works only when the flat list is divided evenly. If the last row has partial elements, we'll need to process them separately.

We can use this kind of function to group data into same-sized tuples, with an odd-sized tuple at the end, using the following definitions:

```
ItemType = TypeVar("ItemType")
Flat = Sequence[ItemType]
Grouped = List[Tuple[ItemType, ...]]

def group_by_seq(n: int, sequence: Flat) -> Grouped:
```

```
    flat_iter=iter(sequence)
    full_sized_items = list( tuple(next(flat_iter)
        for i in range(n))
            for row in range(len(sequence)//n))
    trailer = tuple(flat_iter)
    if trailer:
        return full_sized_items + [trailer]
    else:
        return full_sized_items
```

Within the `group_by_seq()` function, an initial `list` is built and assigned to the variable `full_sized_items`. Each `tuple` in this list is of size `n`. If there are leftovers, the trailing items are used to build a `tuple` with a non-zero length that we can append to the `list` of full-sized items. If the trailer `tuple` is of the length zero, it can be safely ignored.

The type hints include a generic definition of `ItemType` as a type variable. The intent of a type variable is to show that whatever type is an input to this function will be returned from the function. A sequence of strings or a sequence of floats would both work properly.

The input is summarized as a `Sequence` of items. The output is a `List` of `Tuples` of items. The items are all of a common type, described with the `ItemType` type variable.

This isn't as delightfully simple and functional-looking as other algorithms we've looked at. We can rework this into a simpler generator function that yields an iterable instead of a list.

The following code uses a `while` loop as part of tail-recursion optimization:

```
ItemType = TypeVar("ItemType")
Flat_Iter = Iterator[ItemType]
Grouped_Iter = Iterator[Tuple[ItemType, ...]]

def group_by_iter(n: int, iterable: Flat_Iter) -> Grouped_Iter:
    row = tuple(next(iterable) for i in range(n))
    while row:
        yield row
        row = tuple(next(iterable) for i in range(n))
```

We've created a row of the required length from the input iterable. At the end of the input iterable, the value of `tuple(next(iterable) for i in range(n))` will be a zero-length tuple. This can be the base case of a recursive definition. This was manually optimized into the terminating condition of the `while` statement.

The type hints have been modified to reflect the way this works with an iterator. It is not limited to sequences. Because it uses `next()` explicitly, it has to be used like this: `group_by_iter(7, iter(flat))`. The `iter()` function must be used to create an iterator from a collection.

Structuring flat sequences – an alternative approach

Let's say we have a simple, flat `list` and we want to create pairs from this list. The following is the required data:

```
flat= ['2', '3', '5', '7', '11', '13', '17', '19', '23', '29',
 '31', '37', '41', '43', '47', '53', '59', '61', '67', '71',... ]
```

We can create pairs using list slices, as follows:

```
zip(flat[0::2], flat[1::2])
```

The slice `flat[0::2]` is all of the even positions. The slice `flat[1::2]` is all of the odd positions. If we zip these together, we get a two-tuple. The item at index `[0]` is the value from the first even position, and then the item at index `[1]` is the value from the first odd position. If the number of elements is even, this will produce pairs nicely. If the total number of items is odd, the item will be dropped; there's a handy solution to this.

This expression has the advantage of being quite short. The functions shown in the previous section are longer ways to solve the same problem.

This approach can be generalized. We can use the `*(args)` approach to generate a sequence-of-sequences that must be zipped together. It looks like the following:

```
zip(*(flat[i::n] for i in range(n)))
```

This will generate `n` slices—`flat[0::n]`, `flat[1::n]`, `flat[2::n]`, and so on, and `flat[n-1::n]`. This collection of slices becomes the arguments to `zip()`, which then interleaves values from each slice.

Recall that `zip()` truncates the sequence at the shortest `list`. This means that if the `list` is not an even multiple of the grouping factor `n`, (`len(flat)%n != 0`), which is the final slice, it won't be the same length as the others, and the others will all be truncated. This is rarely what we want.

If we use the `itertools.zip_longest()` method, then we'll see that the final tuple will be padded with enough `None` values to make it have a length of `n`. In some cases, this padding is acceptable. In other cases, the extra values are undesirable.

The `list` slicing approach to grouping data is another way to approach the problem of structuring a flat sequence of data into blocks. As it is a general solution, it doesn't seem to offer too many advantages over the functions in the previous section. As a solution specialized for making two-tuples from a flat last, it's elegantly simple.

Using reversed() to change the order

There are times when we need a sequence reversed. Python offers us two approaches to this: the `reversed()` function, and slices with reversed indices.

For example, consider performing a base conversion to hexadecimal or binary. The following code is a simple conversion function:

```
def digits(x: int, b: int) -> Iterator[int]:
    if x == 0: return
    yield x % b
    for d in digits(x//b, b):
        yield d
```

This function uses a recursion to yield the digits from the least significant to the most significant. The value of `x%b` will be the least significant digits of `x` in the base `b`.

We can formalize it as follows:

$$\text{digits}(x, b) = \begin{cases} [\,] & \textbf{if } x = 0 \\ [x \mod b] + \text{digits}(\frac{x}{b}, b) & \textbf{if } x > 0 \end{cases}$$

In many cases, we'd prefer the digits to be yielded in the reverse order. We can wrap this function with the `reversed()` function to swap the order of the digits:

```
def to_base(x: int, b: int) -> Iterator[int]:
    return reversed(tuple(digits(x, b)))
```

The `reversed()` function produces an iterable, but the argument value must be a sequence object. The function then yields the items from that object in the reverse order.

We can do a similar kind of thing with a slice, such as `tuple(digits(x, b))[::-1]`. The slice, however, is not an iterator. A slice is a materialized object built from another materialized object. In this case, for such small collections of values, the distinction is minor. As the `reversed()` function uses less memory, it can be advantageous for larger collections.

Using enumerate() to include a sequence number

Python offers the `enumerate()` function to apply index information to values in a sequence or iterable. It performs a specialized kind of wrap that can be used as part of an `unwrap(process(wrap(data)))` design pattern.

It looks like the following code snippet:

```
>>> xi
[1.47, 1.5, 1.52, 1.55, 1.57, 1.6, 1.63, 1.65, 1.68, 1.7, 1.73,
 1.75, 1.78, 1.8, 1.83]
>>> list(enumerate(xi))
[(0, 1.47), (1, 1.5), (2, 1.52), (3, 1.55), (4, 1.57),
 (5, 1.6), (6, 1.63), (7, 1.65), (8, 1.68), (9, 1.7),
 (10, 1.73), (11, 1.75), (12, 1.78), (13, 1.8), (14, 1.83)]
```

The `enumerate()` function transformed each input `item` into a pair with a sequence number and the original `item`. It's vaguely similar to the following:

```
zip(range(len(source)), source)
```

An important feature of `enumerate()` is that the result is an iterable and it works with any iterable input.

When looking at statistical processing, for example, the `enumerate()` function comes in handy to transform a single sequence of values into a more proper time series by prefixing each sample with a number.

Summary

In this chapter, we saw detailed ways to use a number of built-in reductions.

We've used `any()` and `all()` to do essential logic processing. These are tidy examples of reductions using a simple operator, such as `or` or `and`.

We've also looked at numeric reductions such as, `len()` and `sum()`. We've applied these functions to create some higher-order statistical processing. We'll return to these reductions in `Chapter 6`, *Recursions and Reductions*.

We've also looked at some of the built-in mappings.

The `zip()` function merges multiple sequences. This leads us to look at using this in the context of structuring and flattening more complex data structures. As we'll see in examples in later chapters, nested data is helpful in some situations and flat data is helpful in others.

The `enumerate()` function maps an iterable to a sequence of two-tuples. Each two-tuple has the sequence number at index `[0]` and the original value at index `[1]`.

The `reversed()` function iterates over the items in a sequence object, with their original order reversed. Some algorithms are more efficient at producing results in one order, but we'd like to present these results in the opposite order.

In the next chapter, we'll look at the mapping and reduction functions that use an additional function as an argument to customize their processing. Functions that accept a function as an argument are our first examples of higher-order functions. We'll also touch on functions that return functions as a result.

5
Higher-Order Functions

A very important feature of the functional programming paradigm is higher-order functions. These are functions that accept functions as arguments or return functions as results. Python offers several kinds of higher-order functions. We'll look at them and some logical extensions.

As we can see, there are three varieties of higher-order functions as follows:

- Functions that accept functions as one (or more) of their arguments
- Functions that return a function
- Functions that accept a function and return a function, a combination of the preceding two features

Python offers several higher-order functions of the first variety. We'll look at these built-in higher-order functions in this chapter. We'll look at a few of the library modules that offer higher-order functions in later chapters.

The idea of a function that emits functions can seem a bit odd. However, when we look at a `Callable` class, the class definition is a function that returns `Callable` objects when evaluated. This is one example of a function that creates another function.

Functions that accept functions and create functions include complex callable classes as well as function decorators. We'll introduce decorators in this chapter, but defer deeper consideration of decorators until `Chapter 11`, *Decorator Design Techniques*.

Sometimes, we wish that Python had higher-order versions of the collection functions from the previous chapter. In this chapter, we'll show the `reduce(extract())` design pattern to perform a reduction of specific fields extracted from a larger tuple. We'll also look at defining our own version of these common collection-processing functions.

In this chapter, we'll look at the following functions:

- `max()` and `min()`
- `map()`
- `filter()`
- `iter()`
- `sorted()`

We'll also look at lambda forms that we can use to simplify using higher-order functions.

There are a number of higher-order functions in the `itertools` module. We'll look at this module in `Chapter 8`, *The Itertools Module* and `Chapter 9`, *More Itertools Techniques*.

Additionally, the `functools` module provides a general-purpose `reduce()` function. We'll look at this in `Chapter 10`, *The Functools Module* because it's not as generally applicable as the other higher-order functions in this chapter.

The `max()` and `min()` functions are reductions; they create a single value from a collection. The other functions are mappings. They don't reduce the input to a single value.

The `max()`, `min()`, and `sorted()` functions have both a default behavior as well as a higher-order function behavior. A function can be provided via the `key=` argument. The `map()` and `filter()` functions take the function as the first positional argument.

Using max() and min() to find extrema

The `max()` and `min()` functions each have a dual life. They are simple functions that apply to collections. They are also higher-order functions. We can see their default behavior as follows:

```
>>> max(1, 2, 3)
3
>>> max((1,2,3,4))
4
```

Both functions will accept an indefinite number of arguments. The functions are designed to also accept a sequence or an iterable as the only argument and locate `max` (or `min`) of that iterable.

They also do something more sophisticated. Let's say we have our trip data from the examples in `Chapter 4`, *Working with Collections*. We have a function that will generate a sequence of tuples that looks as follows:

```
(
 ((37.54901619777347, -76.33029518659048), (37.840832, -76.273834),
17.7246),
 ((37.840832, -76.273834), (38.331501, -76.459503), 30.7382),
 ((38.331501, -76.459503), (38.845501, -76.537331), 31.0756),
 ((36.843334, -76.298668), (37.549, -76.331169), 42.3962),
 ((37.549, -76.331169), (38.330166, -76.458504), 47.2866),
 ((38.330166, -76.458504), (38.976334, -76.473503), 38.8019)
)
```

Each `tuple` in this collection has three values: a starting location, an ending location, and a distance. The locations are given in latitude and longitude pairs. The east latitude is positive, so these are points along the US East Coast, about 76° west. The distances between points are in nautical miles.

We have three ways of getting the maximum and minimum distances from this sequence of values. They are as follows:

- Extract the distance with a generator function. This will give us only the distances, as we've discarded the other two attributes of each leg. This won't work out well if we have any additional processing requirements.
- Use the `unwrap(process(wrap()))` pattern. This will give us the legs with the longest and shortest distances. From these, we can extract the distance item, if that's all that's needed.
- Use the `max()` and `min()` functions as higher-order functions, inserting a function that does the extraction of the important distance values.

To provide context, the following is a script that builds the overall trip:

```
from ch02_ex3 import (
    float_from_pair, lat_lon_kml, limits, haversine, legs
)
path = float_from_pair(float_lat_lon(row_iter_kml(source)))
trip = tuple(
    (start, end, round(haversine(start, end), 4))
        for start, end in legs(iter(path)))
```

This script requires `source` to be an open file with KML-formatted data points. The essential `trip` object is a `tuple` of individual legs. Each leg is a three-tuple with the starting point, the ending point, and the distance, computed with the `haversine` function. The `leg` function creates start-end pairs from the overall `path` of points in the original KML file.

Once we have this `trip` object, we can extract distances and compute the maximum and minimum of those distances. The code to do this with a generator function looks as follows:

```
>>> long = max(dist for start, end, dist in trip)
>>> short = min(dist for start, end, dist in trip)
```

We've used a generator function to extract the relevant item from each leg of the `trip` tuple. We've had to repeat the generator function because each generator expression can be consumed only once.

Here are the results based on a larger set of data than was shown previously:

```
>>> long
129.7748
>>> short
0.1731
```

The following is a version with the `unwrap(process(wrap()))` pattern. To make it clear, the example includes functions with the names `wrap()` and `unwrap()`. Here's the functions and the evaluation of those functions:

```
from typing import Iterator, Iterable, Tuple, Any

Wrapped = Tuple[Any, Tuple]
def wrap(leg_iter: Iterable[Tuple]) -> Iterable[Wrapped]:
    return ((leg[2], leg) for leg in leg_iter)
def unwrap(dist_leg: Tuple[Any, Any]) -> Any:
    distance, leg = dist_leg
    return leg

long = unwrap(max(wrap(trip)))
short = unwrap(min(wrap(trip)))
```

Unlike the previous version, the `max()` and `min()` functions will locate all attributes of the `legs` with the longest and shortest distances. Rather than simply extracting the distances, we put the distances first in each wrapped tuple. We can then use the default forms of the `min()` and `max()` functions to process the two tuples that contain the distance and leg details. After processing, we can strip the first element, leaving just the `leg` details.

The results look as follows:

```
((27.154167, -80.195663), (29.195168, -81.002998), 129.7748)
((35.505665, -76.653664), (35.508335, -76.654999), 0.1731)
```

The final and most important form uses the higher-order function feature of the `max()` and `min()` functions. We'll define a `helper` function first and then use it to reduce the collection of legs to the desired summaries by executing the following code snippet:

```
def by_dist(leg: Tuple[Any, Any, Any]) -> Any:
    lat, lon, dist = leg
    return dist

long = max(trip, key=by_dist)
short = min(trip, key=by_dist)
```

The `by_dist()` function picks apart the three items in each `leg` tuple and returns the distance item. We'll use this with the `max()` and `min()` functions.

The `max()` and `min()` functions both accept an iterate and a function as arguments. The keyword parameter `key=` is used by all of Python's higher-order functions to provide a function that will be used to extract the necessary key value.

We can use the following to help conceptualize how the `max()` function uses the `key` function:

```
from typing import Iterable, Any, Callable

def max_like(trip: Iterable[Any], key: Callable) -> Any:
    wrap = ((key(leg), leg) for leg in trip)
    return sorted(wrap)[-1][1]
```

The `max()` and `min()` functions behave as if the result of the given `key()` function is being used to wrap each item in the sequence in a two-tuple. After sorting the two-tuples, selecting the first (for `min`) or last (for `max`) provides a two-tuple with an extreme value. This can be decomposed to recover the original value.

In order for the `key()` function to be optional, it can have a default value of `lambda x: x`.

Using Python lambda forms

In many cases, the definition of a helper function seems to requires too much code. Often, we can digest the `key` function to a single expression. It can seem wasteful to have to write both `def` and `return` statements to wrap a single expression.

Python offers the lambda form as a way to simplify using higher-order functions. A lambda form allows us to define a small, anonymous function. The function's body is limited to a single expression.

The following is an example of using a simple `lambda` expression as the key:

```
long = max(trip, key=lambda leg: leg[2])
short = min(trip, key=lambda leg: leg[2])
```

The `lambda` we've used will be given an item from the sequence; in this case, each leg three tuple will be given to the `lambda`. The `lambda` argument variable, `leg`, is assigned and the expression, `leg[2]`, is evaluated, plucking the distance from the three tuple.

In cases where a lambda is used exactly once, this form is ideal. When reusing a lambda, it's important to avoid copy and paste. What's the alternative?

We can assign lambdas to variables, by doing something like this:

```
start = lambda x: x[0]
end = lambda x: x[1]
dist = lambda x: x[2]
```

Each of these `lambda` forms is a callable object, similar to a defined function. They can be used like a function.

The following is an example at the interactive prompt:

```
>>> leg = ((27.154167, -80.195663), (29.195168, -81.002998), 129.7748)
>>> start = lambda x: x[0]
>>> end = lambda x: x[1]
>>> dist = lambda x: x[2]
>>> dist(leg)
129.7748
```

Python offers us two ways to assign meaningful names to elements of tuples: namedtuples and a collection of lambdas. Both are equivalent. We can use lambdas instead of namedtuples.

To extend this example, we'll look at how we get the `latitude` or `longitude` value of the starting or ending point. This is done by defining some additional lambdas.

The following is a continuation of the interactive session:

```
>>> start(leg)
(27.154167, -80.195663)

>>> lat = lambda x: x[0]
>>> lon = lambda x: x[1]
>>> lat(start(leg))
27.154167
```

There's no clear advantage to using lambdas as a way to extract fields over namedtuples. A set of lambda objects to extract fields requires more lines of code to define than a namedtuple. On the other hand, lambdas allow the code to rely on prefix function notation, which might be easier to read in a functional programming context. More importantly, as we'll see in the `sorted()` example later, `lambdas` can be used more effectively than `namedtuple` attribute names with `sorted()`, `min()`, and `max()`.

Lambdas and the lambda calculus

In a book on a purely functional programming language, it would be necessary to explain lambda calculus, and the technique invented by Haskell Curry that we call **currying**. Python, however, doesn't stick closely to this kind of lambda calculus. Functions are not curried to reduce them to single-argument lambda forms.

Python lambda forms are not restricted to single argument functions. They can have any number of arguments. They are restricted to a single expression, however.

We can, using the `functools.partial` function, implement currying. We'll save this for `Chapter 10`, *The Functools Module*.

Using the map() function to apply a function to a collection

A scalar function maps values from a domain to a range. When we look at the `math.sqrt()` function, as an example, we're looking at a mapping from the `float` value, `x`, to another `float` value, `y = sqrt(x)`, such that $y^2 = x$. The domain is limited to positive values. The mapping can be done via a calculation or table interpolation.

The `map()` function expresses a similar concept; it maps values from one collection to create another collection. It assures that the given function is used to map each individual item from the domain collection to the range collection-the ideal way to apply a built-in function to a collection of data.

Our first example involves parsing a block of text to get the sequence of numbers. Let's say we have the following chunk of text:

```
>>> text= """\
...    2   3    5    7   11   13   17   19   23   29
...   31  37   41   43   47   53   59   61   67   71
...   73  79   83   89   97  101  103  107  109  113
...  127 131  137  139  149  151  157  163  167  173
...  179 181  191  193  197  199  211  223  227  229
... """
```

We can restructure this text using the following generator function:

```
>>> data= list(
...     v for line in text.splitlines()
...         for v in line.split())
```

This will split the text into lines. For each line, it will split the line into space-delimited words and iterate through each of the resulting strings. The results look as follows:

```
['2', '3', '5', '7', '11', '13', '17', '19', '23', '29',
 '31', '37', '41', '43', '47', '53', '59', '61', '67', '71',
 '73', '79', '83', '89', '97', '101', '103', '107', '109', '113',
 '127', '131', '137', '139', '149', '151', '157', '163', '167',
 '173', '179', '181', '191', '193', '197', '199', '211', '223',
 '227', '229']
```

We still need to apply the `int()` function to each of the `string` values. This is where the `map()` function excels. Take a look at the following code snippet:

```
>>> list(map(int, data))
[2, 3, 5, 7, 11, 13, 17, 19, 23, 29, 31, 37, 41, 43, 47, 53, 59,
 61, 67, 71, 73, 79, 83, 89, 97, 101, 103, 107, 109, 113, 127, 131,
 137, 139, 149, 151, 157, 163, 167, 173, 179, 181, 191, 193, 197,
 199, 211, 223, 227, 229]
```

The `map()` function applied the `int()` function to each value in the collection. The result is a sequence of numbers instead of a sequence of strings.

The `map()` function's results are iterable. The `map()` function can process any type of iterable.

The idea here is that any Python function can be applied to the items of a collection using the `map()` function. There are a lot of built-in functions that can be used in this map-processing context.

Working with lambda forms and map()

Let's say we want to convert our trip distances from nautical miles to statute miles. We want to multiply each leg's distance by 6076.12/5280, which is 1.150780.

We can do this calculation with the `map()` function as follows:

```
map(
    lambda x: (start(x), end(x), dist(x)*6076.12/5280),
    trip
)
```

We've defined a `lambda` that will be applied to each leg in the trip by the `map()` function. The `lambda` will use other `lambdas` to separate the start, end, and distance values from each leg. It will compute a revised distance and assemble a new leg tuple from the start, end, and statute mile distances.

This is precisely like the following generator expression:

```
((start(x), end(x), dist(x)*6076.12/5280) for x in trip)
```

We've done the same processing on each item in the generator expression.

The important difference between the `map()` function and a generator expression is that the `map()` function can have a reusable lambda or function definition. The better approach is to do the following:

```
to_miles = lambda x: start(x), end(x), dist(x)*6076.12/5280
trip_m = map(to_miles, trip)
```

This variant separates the transformation, `to_miles`, from the process applying the transformation to the data.

Using map() with multiple sequences

Sometimes, we'll have two collections of data that need to be parallel to each other. In `Chapter 4`, *Working with Collections*, we saw how the `zip()` function can interleave two sequences to create a sequence of pairs. In many cases, we're really trying to do something like the following:

```
map(function, zip(one_iterable, another_iterable))
```

We're creating argument tuples from two (or more) parallel iterables and applying a function to the argument `tuple`. We can also look at it as follows:

```
(function(x,y)
    for x,y in zip(one_iterable, another_iterable)
)
```

Here, we've replaced the `map()` function with an equivalent generator expression.

We might have the idea of generalizing the whole thing to the following:

```
def star_map(function, *iterables)
    return (function(*args) for args in zip(*iterables))
```

There is a better approach that is already available to us. We don't actually need these techniques. Let's look at a concrete example of the alternate approach.

In `Chapter 4`, *Working with Collections*, we looked at trip data that we extracted from an XML file as a series of waypoints. We needed to create legs from this list of waypoints that show the start and end of each leg.

The following is a simplified version that uses the `zip()` function applied to a special kind of iterable:

```
>>> waypoints = range(4)
```

```
>>> zip(waypoints, waypoints[1:])
<zip object at 0x101a38c20>
>>> list(_)
[(0, 1), (1, 2), (2, 3)]
```

We've created a sequence of pairs drawn from a single flat list. Each pair will have two adjacent values. The `zip()` function properly stops when the shorter list is exhausted. This `zip( x, x[1:])` pattern only works for materialized sequences and the iterable created by the `range()` function.

We created pairs so that we can apply the `haversine()` function to each pair to compute the distance between the two points on the path. The following is how it looks in one sequence of steps:

```
from ch02_ex3 import (lat_lon_kml, float_from_pair, haversine)

path = tuple(float_from_pair(lat_lon_kml()))
distances_1 = map(
    lambda s_e: (s_e[0], s_e[1], haversine(*s_e)),
    zip(path, path[1:])
)
```

We've loaded the essential sequence of waypoints into the `path` variable. This is an ordered sequence of latitude-longitude pairs. As we're going to use the `zip(path, path[1:])` design pattern, we must have a materialized sequence and not a simple iterable.

The results of the `zip()` function will be pairs that have a start and end. We want our output to be a triple with the start, end, and distance. The `lambda` we're using will decompose the original start-end two-tuple and create a new three tuple from the start, end, and distance.

As noted previously, we can simplify this by using a clever feature of the `map()` function, which is as follows:

```
distances_2 = map(
    lambda s, e: (s, e, haversine(s, e)),
    path, path[1:])
```

Note that we've provided a function and two iterables to the `map()` function. The `map()` function will take the next item from each iterable and apply those two values as the arguments to the given function. In this case, the given function is a `lambda` that creates the desired three tuple from the start, end, and distance.

The formal definition for the `map()` function states that it will do **star-map** processing with an indefinite number of iterables. It will take items from each iterable to create a tuple of argument values for the given function.

Using the filter() function to pass or reject data

The job of the `filter()` function is to use and apply a decision function called a predicate to each value in a collection. A decision of `True` means that the value is passed; otherwise, the value is rejected. The `itertools` module includes `filterfalse()` as variations on this theme. Refer to `Chapter 8`, *The Itertools Module,* to understand the usage of the `itertools` module's `filterfalse()` function.

We might apply this to our trip data to create a subset of legs that are over 50 nautical miles long, as follows:

```
long= list(
    filter(lambda leg: dist(leg) >= 50, trip))
)
```

The predicate `lambda` will be `True` for long legs, which will be passed. Short legs will be rejected. The output is the 14 legs that pass this distance test.

This kind of processing clearly segregates the filter rule (`lambda leg: dist(leg) >= 50`) from any other processing that creates the `trip` object or analyzes the long legs.

For another simple example, look at the following code snippet:

```
>>> filter(lambda x: x%3==0 or x%5==0, range(10))
<filter object at 0x101d5de50>
>>> sum(_)
23
```

We've defined a simple `lambda` to check whether a number is a multiple of three or a multiple of five. We've applied that function to an iterable, `range(10)`. The result is an iterable sequence of numbers that are passed by the decision rule.

The numbers for which the `lambda` is `True` are `[0, 3, 5, 6, 9]`, so these values are passed. As the `lambda` is `False` for all other numbers, they are rejected.

This can also be done with a generator expression by executing the following code:

```
>>> list(x for x in range(10) if x%3==0 or x%5==0)
[0, 3, 5, 6, 9]
```

We can formalize this using the following set comprehension notation:

$$\{x|0 \leq x < 10 \wedge (x \mod 3 = 0 \vee x \mod 5 = 0)\}$$

This says that we're building a collection of *x* values such that *x* is in `range(10)` and `x%3==0 or x%5==0`. There's a very elegant symmetry between the `filter()` function and formal mathematical set comprehensions.

We often want to use the `filter()` function with defined functions instead of lambda forms. The following is an example of reusing a predicate defined earlier:

```
>>> from ch01_ex1 import isprimeg
>>> list(filter(isprimeg, range(100)))
[2, 3, 5, 7, 11, 13, 17, 19, 23, 29, 31, 37, 41, 43, 47, 53, 59, 61, 67,
71, 73, 79, 83, 89, 97]
```

In this example, we imported a function from another module called `isprimeg()`. We then applied this function to a collection of values to pass the prime numbers and rejected any non-prime numbers from the collection.

This can be a remarkably inefficient way to generate a table of prime numbers. The superficial simplicity of this is the kind of thing lawyers call an *attractive nuisance*. It looks like it might be fun, but it doesn't scale well at all. The `isprimeg()` function duplicates all of the testing effort for each new value. Some kind of cache is essential to provide redoing the testing of primality. A better algorithm is the **Sieve of Eratosthenes**; this algorithm retains the previously located prime numbers and uses them to prevent recalculation.

Using filter() to identify outliers

In the previous chapter, we defined some useful statistical functions to compute mean and standard deviation and normalize a value. We can use these functions to locate outliers in our trip data. What we can do is apply the `mean()` and `stdev()` functions to the distance value in each leg of a trip to get the population mean and standard deviation.

We can then use the `z()` function to compute a normalized value for each leg. If the normalized value is more than 3, the data is extremely far from the mean. If we reject these outliers, we have a more uniform set of data that's less likely to harbor reporting or measurement errors.

The following is how we can tackle this:

```
from stats import mean, stdev, z

dist_data = list(map(dist, trip))
μ_d = mean(dist_data)
σ_d = stdev(dist_data)
outlier = lambda leg: z(dist(leg), μ_d, σ_d) > 3
print("Outliers", list(filter(outlier, trip)))
```

We've mapped the distance function to each `leg` in the `trip` collection. As we'll do several things with the result, we must materialize a `list` object. We can't rely on the iterator as the first function will consume it. We can then use this extraction to compute population statistics `μ_d` and `σ_d` with the mean and standard deviation.

Given the statistics, we used the outlier lambda to `filter` our data. If the normalized value is too large, the data is an outlier.

The result of `list(filter(outlier, trip))` is a list of two legs that are quite long compared to the rest of the legs in the population. The average distance is about 34 nm, with a standard deviation of 24 nm. No trip can have a normalized distance of less than -1.407.

We're able to decompose a fairly complex problem into a number of independent functions, each one of which can be easily tested in isolation. Our processing is a composition of simpler functions. This can lead to succinct, expressive functional programming.

The iter() function with a sentinel value

The built-in `iter()` function creates an iterator over an object of one of the collection classes. The `list`, `dict`, and `set` classes all work with the `iter()` function to provide an `iterator` object for the items in the underlying collection. In most cases, we'll allow the `for` statement to do this implicitly. In a few cases, however, we need to create an iterator explicitly. One example of this is to separate the head from the tail of a collection.

Other uses include building iterators to consume the values created by a callable object (for example, a function) until a sentinel value is found. This feature is sometimes used with the `read()` function of a file to consume items until some end-of-line or end-of-file sentinel value is found. An expression such as `iter(file.read, '\n')` will evaluate the given function until the sentinel value, `'\n'`, is found. This must be used carefully: if the sentinel is not found, it can continue reading zero-length strings forever.

Providing a `callable` function to `iter()` can be a bit challenging because the function we provide must maintain some state internally. This is generally looked at as undesirable in functional programs. However, hidden state is a feature of an open file, for example, each `read()` or `readline()` function advances the internal state to the next character or next line.

Another example of explicit iteration is the way that a mutable collection object's `pop()` method makes a stateful change to a collection object. The following is an example of using the `pop()` method:

```
>>> tail = iter([1, 2, 3, None, 4, 5, 6].pop, None)
>>> list(tail)
[6, 5, 4]
```

The `tail` variable was set to an iterator over the list `[1, 2, 3, None, 4, 5, 6]` that will be traversed by the `pop()` function. The default behavior of `pop()` is `pop(-1)`, that is, the elements are popped in the reverse order. This makes a stateful change to the list object: each time `pop()` is called, the item is removed, mutating the list. When the `sentinel` value is found, the `iterator` stops returning values. If the sentinel is not found, this will break with an `IndexError` exception.

This kind of internal state management is something we'd like to avoid. Consequently, we won't try to contrive a use for this feature.

Using sorted() to put data in order

When we need to produce results in a defined order, Python gives us two choices. We can create a `list` object and use the `list.sort()` method to put items in an order. An alternative is to use the `sorted()` function. This function works with any iterable, but it creates a final `list` object as part of the sorting operation.

The sorted() function can be used in two ways. It can be simply applied to collections. It can also be used as a higher-order function using the key= argument.

Let's say we have our trip data from the examples in Chapter 4, *Working with Collections*. We have a function that will generate a sequence of tuples with start, end, and distance for each leg of a trip. The data looks as follows:

```
(
  ((37.54901619777347, -76.33029518659048), (37.840832, -76.273834),
17.7246),
  ((37.840832, -76.273834), (38.331501, -76.459503), 30.7382),
  ((38.331501, -76.459503), (38.845501, -76.537331), 31.0756),
  ((36.843334, -76.298668), (37.549, -76.331169), 42.3962),
  ((37.549, -76.331169), (38.330166, -76.458504), 47.2866),
  ((38.330166, -76.458504), (38.976334, -76.473503), 38.8019)
)
```

We can see the default behavior of the sorted() function using the following interaction:

```
>>> sorted(dist(x) for x in trip)
[0.1731, 0.1898, 1.4235, 4.3155, ... 86.2095, 115.1751, 129.7748]
```

We used a generator expression (dist(x) for x in trip) to extract the distances from our trip data. We then sorted this iterable collection of numbers to get the distances from 0.17 nm to 129.77 nm.

If we want to keep the legs and distances together in their original three tuples, we can have the sorted() function apply a key() function to determine how to sort the tuples, as shown in the following code snippet:

```
>>> sorted(trip, key=dist)
[
  ((35.505665, -76.653664), (35.508335, -76.654999), 0.1731),
  ((35.028175, -76.682495), (35.031334, -76.682663), 0.1898),
  ((27.154167, -80.195663), (29.195168, -81.002998), 129.7748)
]
```

We've sorted the trip data, using a distlambda to extract the distance from each tuple. The dist function is simply as follows:

```
dist = lambda leg: leg[2]
```

This shows the power of using simple lambda to decompose a complex tuple into its constituent elements.

Writing higher-order functions

We can identify three varieties of higher-order function; they are as follows:

- Functions that accept a function as one of their arguments.
- Functions that return a function. A `Callable` class is a common example of this. A function that returns a generator expression can be thought of as a higher-order function.
- Functions that accept and return a function. The `functools.partial()` function is a common example of this. We'll save this for `Chapter 10`, *The Functools Module*. A decorator is different; we'll save this for `Chapter 11`, *Decorator Design* Techniques.

We'll expand on these simple patterns using a higher-order function to also transform the structure of the data. We can do several common transformations, such as the following:

- Wrap objects to create more complex objects
- Unwrap complex objects into their components
- Flatten a structure
- Structure a flat sequence

A `Callable` class object is a commonly used example of a function that returns a `callable` object. We'll look at this as a way to write flexible functions into which configuration parameters can be injected.

We'll also introduce simple decorators in this chapter. We'll defer deeper consideration of decorators until `Chapter 11`, *Decorator Design Techniques*.

Writing higher-order mappings and filters

Python's two built-in higher-order functions, `map()` and `filter()`, generally handle almost everything we might want to throw at them. It's difficult to optimize them in a general way to achieve higher performance. We'll look at functions of Python 3.4, such as `imap()`, `ifilter()`, and `ifilterfalse()`, in `Chapter 8`, *The Itertools Module*.

We have three, largely equivalent ways to express a mapping. Assume that we have some function, `f(x)`, and some collection of objects, `C`. We have three entirely equivalent ways to express a mapping; they are as follows:

- The `map()` function:

```
map(f, C)
```

- A generator expression:

```
(f(x) for x in C)
```

- A generator function with a `yield` statement:

```
def mymap(f, C):
    for x in C:
        yield f(x)
mymap(f, C)
```

Similarly, we have three ways to apply a `filter` function to a `collection`, all of which are equivalent:

- The `filter()` function:

```
filter(f, C)
```

- A generator expression:

```
(x for x in C if f(x))
```

- A generator function with a `yield` statement:

```
def myfilter(f, C):
    for x in C:
        if f(x):
            yield x
myfilter(f, C)
```

There are some performance differences; often the `map()` and `filter()` functions are fastest. More importantly, there are different kinds of extensions that fit these mapping and filtering designs, which are as follows:

- We can create a more sophisticated function, `g(x)`, that is applied to each element, or we can apply a function to the whole collection prior to processing. This is the most general approach and applies to all three designs. This is where the bulk of our functional design energy is invested.

- We can tweak the `for` loop inside the generator expression or generator function. One obvious tweak is to combine mapping and filtering into a single operation by extending the generator expression with an `if` clause. We can also merge the `mymap()` and `myfilter()` functions to combine mapping and filtering.

Profound changes that alter the structure of the data handled by the loop often happen as software evolves and matures. We have a number of design patterns, including wrapping, unwrapping (or extracting), flattening, and structuring. We've looked at a few of these techniques in previous chapters.

We need to exercise some caution when designing mappings that combine too many transformations in a single function. As far as possible, we want to avoid creating functions that fail to be succinct or expressive of a single idea. As Python doesn't have an optimizing compiler, we might be forced to manually optimize slow applications by combining functions. We need to do this kind of optimization reluctantly, only after profiling a poorly performing program.

Unwrapping data while mapping

When we use a construct, such as `(f(x) for x, y in C)`, we use multiple assignments in the `for` statement to unwrap a multi-valued tuple and then apply a function. The whole expression is a mapping. This is a common Python optimization to change the structure and apply a function.

We'll use our trip data from `Chapter 4`, *Working with Collections*. The following is a concrete example of unwrapping while mapping:

```
from typing import Callable, Iterable, Tuple, Iterator, Any

Conv_F = Callable[[float], float]
Leg = Tuple[Any, Any, float]

def convert(
        conversion: Conv_F,
        trip: Iterable[Leg]) -> Iterator[float]:
    return (
        conversion(distance) for start, end, distance in trip
    )
```

This higher-order function would be supported by conversion functions that we can apply to our raw data, as follows:

```
to_miles = lambda nm: nm*5280/6076.12
to_km = lambda nm: nm*1.852
to_nm = lambda nm: nm
```

This function would then be used, as follows, to extract distance and apply a conversion function:

```
convert(to_miles, trip)
```

As we're unwrapping, the result will be a sequence of `floating-point` values. The results are as follows:

```
[20.397120559090908, 35.37291511060606, ..., 44.652462240151515]
```

This `convert()` function is highly specific to our start-end-distance trip data structure, as the `for` loop decomposes that three tuple.

We can build a more general solution for this kind of unwrapping while mapping a design pattern. It suffers from being a bit more complex. First, we need general-purpose decomposition functions, as in the following code snippet:

```
fst = lambda x: x[0]
snd = lambda x: x[1]
sel2 = lambda x: x[2]
```

We'd like to be able to express `f(sel2(s_e_d)) for s_e_d in trip`. This involves functional composition; we're combining a function, such as `to_miles()`, and a selector, such as `sel2()`. We can express functional composition in Python using yet another lambda, as follows:

```
to_miles = lambda s_e_d: to_miles(sel2(s_e_d))
```

This gives us a longer but more general version of unwrapping, as follows:

```
(to_miles(s_e_d) for s_e_d in trip)
```

While this second version is somewhat more general, it doesn't seem wonderfully helpful.

What's important to note about our higher-order `convert()` function is that we're accepting a function as an argument and returning a generator function as a result. The `convert()` function is not a generator function; it doesn't `yield` anything. The result of the `convert()` function is a generator expression that must be evaluated to accumulate the individual values. We've used `Iterator[float]` to emphasize that the result is an iterator; a subclass of Python's generator functions.

The same design principle works to create hybrid filters instead of mappings. We'd apply the filter in an `if` clause of the generator expression that was returned.

We can combine mapping and filtering to create yet more complex functions. While it is appealing to create more complex functions, it isn't always valuable. A complex function might not beat the performance of a nested use of simple `map()` and `filter()` functions. Generally, we only want to create a more complex function if it encapsulates a concept and makes the software easier to understand.

Wrapping additional data while mapping

When we use a construct such as `((f(x), x) for x in C)`, we've used wrapping to create a multi-valued tuple while also applying a mapping. This is a common technique to save derived results to create constructs that have the benefits of avoiding recalculation without the liability of complex state-changing objects.

This is part of the example shown in `Chapter 4`, *Working with Collections,* to create the trip data from the path of points. The code looks like this:

```
from ch02_ex3 import (
    float_from_pair, lat_lon_kml, limits, haversine, legs
)

path = float_from_pair(float_lat_lon(row_iter_kml(source)))
trip = tuple(
     (start, end, round(haversine(start, end), 4))
     for start, end in legs(iter(path))
)
```

We can revise this slightly to create a higher-order function that separates the `wrapping` from the other functions. We can define a function as follows:

```
from typing import Callable, Iterable, Tuple, Iterator

Point = Tuple[float, float]
Leg_Raw = Tuple[Point, Point]
```

```
Point_Func = Callable[[Point, Point], float]
Leg_D = Tuple[Point, Point, float]

def cons_distance(
        distance: Point_Func,
        legs_iter: Iterable[Leg_Raw]) -> Iterator[Leg_D]:
    return (
        (start, end, round(distance(start,end), 4))
        for start,end in legs_iter
    )
```

This function will decompose each leg into two variables, `start` and `end`. These variables will be `Point` instances, defined as tuples of two float values. These will be used with the given `distance()` function to compute the distance between the points. The function is a callable that accepts two `Point` objects and returns a `float` result. The result will build a three tuple that includes the original two `Point` objects and also the calculated `float` result.

We can then rewrite our trip assignment to apply the `haversine()` function to compute distances, as follows:

```
path = float_from_pair(float_lat_lon(row_iter_kml(source)))
trip2 = tuple(cons_distance(haversine, legs(iter(path))))
```

We've replaced a generator expression with a higher-order function, `cons_distance()`. The function not only accepts a function as an argument, but it also returns a generator expression.

A slightly different formulation of this is as follows:

```
from typing import Callable, Iterable, Tuple, Iterator, Any
Point = Tuple[float, float]
Leg_Raw = Tuple[Point, Point]
Point_Func = Callable[[Point, Point], float]
Leg_P_D = Tuple[Leg_Raw, ...]

def cons_distance3(
        distance: Point_Func,
        legs_iter: Iterable[Leg_Raw]) -> Iterator[Leg_P_D]:
    return (
        leg + (round(distance(*leg), 4),)  # 1-tuple
        for leg in legs_iter
    )
```

This version makes the construction of a new object built up from an old object a bit clearer. The resulting iterator uses the legs of a trip, defined as `Leg_Raw`, a tuple of two points. It computes the distance along the leg to build the resulting three-tuple with the original `Leg_Raw` object and the distance concatenated to it.

As both of these `cons_distance()` functions accept a function as an argument, we can use this feature to provide an alternative distance formula. For example, we can use the `math.hypot(lat(start)-lat(end), lon(start)-lon(end))` method to compute a less-correct plane distance along each `leg`.

In `Chapter 10`, *The Functools Module*, we'll show how to use the `partial()` function to set a value for the `R` parameter of the `haversine()` function, which changes the units in which the distance is calculated.

Flattening data while mapping

In `Chapter 4`, *Working with Collections*, we looked at algorithms that flattened a nested tuple-of-tuples structure into a single iterable. Our goal at the time was simply to restructure some data, without doing any real processing. We can create hybrid solutions that combine a function with a flattening operation.

Let's assume that we have a block of text that we want to convert to a flat sequence of numbers. The text looks as follows:

```
>>> text= """\
...    2   3   5   7  11  13  17  19  23  29
...   31  37  41  43  47  53  59  61  67  71
...   73  79  83  89  97 101 103 107 109 113
...  127 131 137 139 149 151 157 163 167 173
...  179 181 191 193 197 199 211 223 227 229
... """
```

Each line is a block of 10 numbers. We need to unblock the rows to create a flat sequence of numbers.

This is done with a two-part generator function, as follows:

```
data = list(
    v
    for line in text.splitlines()
    for v in line.split()
)
```

This will split the text into lines and iterate through each line. It will split each line into words and iterate through each word. The output from this is a list of strings, as follows:

```
['2', '3', '5', '7', '11', '13', '17', '19', '23', '29', '31', '37',
 '41', '43', '47', '53', '59', '61', '67', '71', '73', '79', '83',
 '89', '97', '101', '103', '107', '109', '113', '127', '131', '137',
 '139', '149', '151', '157', '163', '167', '173', '179', '181', '191',
 '193', '197', '199', '211', '223', '227', '229']
```

To convert the strings to numbers, we must apply a conversion function as well as unwind the blocked structure from its original format, using the following code snippet:

```
from numbers import Number
from typing import Callable, Iterator

Num_Conv = Callable[[str], Number]

def numbers_from_rows(
        conversion: Num_Conv, text: str) -> Iterator[Number]:
    return (
        conversion(value)
        for line in text.splitlines()
        for value in line.split()
    )
```

This function has a `conversion` argument, which is a function that is applied to each value that will be emitted. The values are created by flattening using the algorithm shown previously.

We can use this `numbers_from_rows()` function in the following kind of expression:

```
print(list(numbers_from_rows(float, text)))
```

Here we've used the built-in `float()` to create a list of `floating-point` values from the block of text.

We have many alternatives using mixtures of higher-order functions and generator expressions. For example, we might express this as follows:

```
map(float,
    value
    for line in text.splitlines()
     for value in line.split()
)
```

This might help us understand the overall structure of the algorithm. The principle is called **chunking**; the details of a function with a meaningful name can be abstracted and we can work with the function in a new context. While we often use higher-order functions, there are times when a generator expression can be clearer.

Structuring data while filtering

The previous three examples combined additional processing with mapping. Combining processing with filtering doesn't seem to be quite as expressive as combining with mapping. We'll look at an example in detail to show that, although it is useful, it doesn't seem to have as compelling a use case as combining mapping and processing.

In `Chapter 4`, *Working with Collections*, we looked at structuring algorithms. We can easily combine a filter with the structuring algorithm into a single, complex function. The following is a version of our preferred function to group the output from an iterable:

```
from typing import Iterator, Tuple

def group_by_iter(n: int, items: Iterator) -> Iterator[Tuple]:
    row = tuple(next(items) for i in range(n))
    while row:
        yield row
        row = tuple(next(items) for i in range(n))
```

This will try to assemble a tuple of `n` items taken from an iterable object. If there are any items in the tuple, they are yielded as part of the resulting iterable. In principle, the function then operates recursively on the remaining items from the original iterable. As the recursion is relatively inefficient in Python, we've optimized it into an explicit `while` loop.

We can use this function as follows:

```
group_by_iter(7,
    filter(lambda x: x%3==0 or x%5==0, range(100))
)
```

This will group the results of applying a `filter()` function to an iterable created by the `range()` function.

We can merge grouping and filtering into a single function that does both operations in a single function body. The modification to `group_by_iter()` looks as follows:

```
def group_filter_iter(
        n: int, pred: Callable, items: Iterator) -> Iterator:
    subset = filter(pred, items)
```

```
    row = tuple(next(subset) for i in range(n))
    while row:
        yield row
        row = tuple(next(subset) for i in range(n))
```

This function applies the filter predicate function to the source iterable provided as the `items` parameter. As the filter output is itself a non-strict iterable, the `subset` value isn't computed in advance; the values are created as needed. The bulk of this function is identical to the version shown previously.

We can slightly simplify the context in which we use this function as follows:

```
group_filter_iter(
    7,
    lambda x: x%3==0 or x%5==0,
    range(1,100)
)
```

Here, we've applied the filter predicate and grouped the results in a single function invocation. In the case of the `filter()` function, it's rarely a clear advantage to apply the filter in conjunction with other processing. It seems as if a separate, visible `filter()` function is more helpful than a combined function.

Writing generator functions

Many functions can be expressed neatly as generator expressions. Indeed, we've seen that almost any kind of mapping or filtering can be done as a generator expression. They can also be done with a built-in higher-order function, such as `map()` or `filter()`, or as a generator function. When considering multiple statement generator functions, we need to be cautious that we don't stray from the guiding principles of functional programming: stateless function evaluation.

Using Python for functional programming means walking on a knife edge between purely functional programming and imperative programming. We need to identify and isolate the places where we must resort to imperative Python code because there isn't a purely functional alternative available.

We're obligated to write generator functions when we need statement features of Python. Features, such as the following, aren't available in generator expressions:

- A `with` context to work with external resources. We'll look at this in `Chapter 6`, *Recursions and Reductions*, where we address file parsing.
- A `while` statement to iterate somewhat more flexibly than a `for` statement. An example of this was shown previously in the *Flattening data while mapping* section.
- A `break` or `return` statement to implement a search that terminates a loop early.
- The `try-except` construct to handle exceptions.
- An internal function definition. We've looked at this in several examples in `Chapter 1`, *Understanding Functional Programming* and `Chapter 2`, *Introducing Essential Functional Concepts*. We'll also revisit it in `Chapter 6`, *Recursions and Reductions*.
- A really complex `if-elif` sequence. Trying to express more than one alternative via `if-else` conditional expressions can become complex looking.
- At the edge of the envelope, we have less-used features of Python, such as `for-else`, `while-else`, `try-else`, and `try-else-finally`. These are all statement-level features that aren't available in generator expressions.

The `break` statement is most commonly used to end processing of a collection early. We can end processing after the first item that satisfies some criteria. This is a version of the `any()` function used to find the existence of a value with a given property. We can also end after processing a larger number of items, but not all of them.

Finding a single value can be expressed succinctly as `min(some-big-expression)` or `max(something big)`. In these cases, we're committed to examining all of the values to assure that we've properly found the minimum or the maximum.

In a few cases, we can stand to have a `first(function, collection)` function where the first value that is `True` is sufficient. We'd like the processing to terminate as early as possible, saving needless calculation.

We can define a function as follows:

```
def first(predicate: Callable, collection: Iterable) -> Any:
    for x in collection:
        if predicate(x): return x
```

We've iterated through the `collection`, applying the given `predicate` function. If the predicate result is `True`, the function returns the associated value and stops processing the iterable. If we exhaust the `collection`, the default value of `None` will be returned.

We can also download a version of this from `PyPi`. The first module contains a variation on this idea. For more details, visit `https://pypi.python.org/pypi/first`.

This can act as a helper when trying to determine whether a number is a prime number or not. The following is a function that tests a number for being prime:

```
import math
def isprimeh(x: int) -> bool:
    if x == 2: return True
    if x % 2 == 0: return False
    factor= first(
        lambda n: x%n==0,
        range(3, int(math.sqrt(x)+.5)+1, 2))
    return factor is None
```

This function handles a few of the edge cases regarding the number `2` being a prime number and every other even number being composite. Then, it uses the `first()` function defined previously to locate the first factor in the given collection.

When the `first()` function will return the factor, the actual number doesn't matter. Its existence is all that matters for this particular example. Therefore, the `isprimeh()` function returns `True` if no factor was found.

We can do something similar to handle data exceptions. The following is a version of the `map()` function that also filters bad data:

```
def map_not_none(func: Callable, source: Iterable) -> Iterator:
    for x in source:
        try:
            yield func(x)
        except Exception as e:
            pass  # For help debugging, use print(e)
```

This function steps through the items in the iterable, assigning each item to the `x` variable. It attempts to apply the function to the item; if no exception is raised, the resulting value is yielded. If an exception is raised, the offending source item is silently dropped.

This can be handy when dealing with data that include values that are not applicable or missing. Rather than working out complex filters to exclude these values, we attempt to process them and drop the ones that aren't valid.

We might use the map() function for mapping not-None values, as follows:

```
data = map_not_none(int, some_source)
```

We'll apply the int() function to each value in some_source. When the some_source parameter is an iterable collection of strings, this can be a handy way to reject strings that don't represent a number.

Building higher-order functions with callables

We can define higher-order functions as callable classes. This builds on the idea of writing generator functions; we'll write callables because we need statement features of Python. In addition to using statements, we can also apply a static configuration when creating the higher-order functions.

What's important about a callable class definition is that the class object, created by the class statement, defines a function that emits a function. Commonly, we'll use a callable object to create a composite function that combines two other functions into something relatively complex.

To emphasize this, consider the following class:

```
from typing import Callable, Optional, Any

class NullAware:
    def __init__(
            self, some_func: Callable[[Any], Any]) -> None:
        self.some_func = some_func
    def __call__(self, arg: Optional[Any]) -> Optional[Any]:
        return None if arg is None else self.some_func(arg)
```

This class is used to create a new function that is nullaware. When an instance of this class is created, a function, some_func, is provided. The only restriction stated is that some_func be Callable[[Any], Any]. This means the argument takes a single argument and results in a single result. The resulting object is callable. A single, optional argument is expected. The implementation of the __call__() method handles the use of None objects as an argument. This method has the effect of making the resulting object Callable[[Optional[Any]], Optional[Any]].

For example, evaluating the NullAware(math.log) expression will create a new function that can be applied to argument values. The __init__() method will save the given function in the resulting object. This object is a function that can then be used to process data.

The common approach is to create the new function and save it for future use by assigning it a name, as follows:

```
null_log_scale = NullAware(math.log)
```

This creates a new function and assigns the name null_log_scale(). We can then use the function in another context. Take a look at the following example:

```
>>> some_data = [10, 100, None, 50, 60]
>>> scaled = map(null_log_scale, some_data)
>>> list(scaled)
[2.302585092994046, 4.605170185988092, None, 3.912023005428146,
4.0943445622221]
```

A less common approach is to create and use the emitted function in one expression, as follows:

```
>>> scaled = map(NullAware(math.log), some_data)
>>> list(scaled)
[2.302585092994046, 4.605170185988092, None, 3.912023005428146,
4.0943445622221]
```

The evaluation of NullAware(math.log) created a function. This anonymous function was then used by the map() function to process an iterable, some_data.

This example's __call__() method relies entirely on expression evaluation. It's an elegant and tidy way to define composite functions built up from lower-level component functions. When working with scalar functions, there are a few complex design considerations. When we work with iterable collections, we have to be a bit more careful.

Assuring good functional design

The idea of stateless functional programming requires some care when using Python objects. Objects are typically stateful. Indeed, one can argue that the entire purpose of object-oriented programming is to encapsulate state change into class definitions. Because of this, we find ourselves pulled in opposing directions between functional programming and imperative programming when using Python class definitions to process collections.

The benefit of using a callable object to create a composite function gives us slightly simpler syntax when the resulting composite function is used. When we start working with iterable mappings or reductions, we have to be aware of how and why we introduce stateful objects.

We'll return to our `sum_filter_f()` composite function shown previously. Here is a version built from a callable class definition:

```
from typing import Callable, Iterable

class Sum_Filter:
    __slots__ = ["filter", "function"]
    def __init__(self,
            filter: Callable[[Any], bool],
            func: Callable[[Any], float]) -> None:
        self.filter = filter
        self.function = func
    def __call__(self, iterable: Iterable) -> float:
        return sum(
            self.function(x)
            for x in iterable
                if self.filter(x)
        )
```

This class has precisely two slots in each object; this puts a few constraints on our ability to use the function as a stateful object. It doesn't prevent all modifications to the resulting object, but it limits us to just two attributes. Attempting to add attributes results in an exception.

The initialization method, `__init__()`, stows the two function names, `filter` and `func`, in the object's instance variables. The `__call__()` method returns a value based on a generator expression that uses the two internal function definitions. The `self.filter()` function is used to pass or reject items. The `self.function()` function is used to transform objects that are passed by the `filter()` function.

An instance of this class is a function that has two strategy functions built into it. We create an instance as follows:

```
count_not_none = Sum_Filter(
    lambda x: x is not None,
    lambda x: 1)
```

We've built a function named `count_not_none()` that counts the `non-None` values in a sequence. It does this by using a `lambda` to pass `non-None` values and a function that uses a constant 1 instead of the actual values present.

Generally, this `count_not_none()` object will behave like any other Python function. The use is somewhat simpler than our previous example of `sum_filter_f()`.

We can use the `count_not_None()` function, as follows:

```
N = count_not_none(data)
```

We would do that instead of using the `sum_filter_f()` function:

```
N = sum_filter_f(valid, count_, data)
```

The `count_not_none()` function, based on a callable, doesn't require quite so many arguments as a conventional function. This makes it superficially simpler to use. However, it can also make it somewhat more obscure because the details of how the function works are in two places in the source code: where the function was created as an instance of a callable class and where the function was used.

Review of some design patterns

The `max()`, `min()`, and `sorted()` functions have a default behavior without a `key=` function. They can be customized by providing a function that defines how to compute a key from the available data. For many of our examples, the `key()` function has been a simple extraction of available data. This isn't a requirement; the `key()` function can do anything.

Imagine the following method: `max(trip, key=random.randint())`. Generally, we try not to have have `key()` functions that do something obscure like this.

The use of a `key=` function is a common design pattern. Functions we design can easily follow this pattern.

We've also looked at lambda forms that we can use to simplify using higher-order functions. One significant advantage of using lambda forms is that it follows the functional paradigm very closely. When writing more conventional functions, we can create imperative programs that might clutter an otherwise succinct and expressive functional design.

We've looked at several kinds of higher-order functions that work with a collection of values. Throughout the previous chapters, we've hinted around at several different design patterns for higher-order functions that apply to collection objects and scalar objects. The following is a broad classification:

- **Return a generator**: A higher-order function can return a generator expression. We consider the function higher-order because it didn't return scalar values or collections of values. Some of these higher-order functions also accept functions as arguments.
- **Act as a generator**: Some function examples use the `yield` statement to make them first-class generator functions. The value of a generator function is an iterable collection of values that are evaluated lazily. We suggest that a generator function is essentially indistinguishable from a function that returns a generator expression. Both are non-strict. Both can yield a sequence of values. For this reason, we'll also consider generator functions as higher order. Built-in functions, such as `map()` and `filter()`, fall into this category.
- **Materialize a collection**: Some functions must return a materialized collection object: `list`, `tuple`, `set`, or `mapping`. These kinds of functions can be of a higher order if they have a function as part of the arguments. Otherwise, they're ordinary functions that happen to work with collections.
- **Reduce a collection**: Some functions work with an iterable (or an object that is some kind of collection) and create a scalar result. The `len()` and `sum()` functions are examples of this. We can create higher-order reductions when we accept a function as an argument. We'll return to this in the next chapter.
- **Scalar**: Some functions act on individual data items. These can be higher-order functions if they accept another function as an argument.

As we design our own software, we can pick and choose among these established design patterns.

Summary

In this chapter, we have seen two reductions that are higher-order functions: `max()` and `min()`. We also looked at the two central higher-order functions, `map()` and `filter()`. We also looked at `sorted()`.

We also looked at how to use a higher-order function to also transform the structure of data. We can perform several common transformations, including wrapping, unwrapping, flattening, and structure sequences of different kinds.

We looked at three ways to define our own higher-order functions, which are as follows:

- The `def` statement. Similar to a `lambda` form that we assign to a variable.
- Defining a callable class as a kind of function that emits composite functions.
- We can also use decorators to emit composite functions. We'll return to this in `Chapter 11`, *Decorator Design Techniques*.

In the next chapter, we'll look at the idea of purely functional iteration via recursion. We'll use Pythonic structures to make several common improvements over purely functional techniques. We'll also look at the associated problem of performing reductions from collections to individual values.

6
Recursions and Reductions

In previous chapters, we've looked at several related kinds of processing designs; some of them are as follows:

- Mapping and filtering, which creates collections from collections
- Reductions that create a scalar value from a collection

The distinction is exemplified by functions such as `map()` and `filter()` that accomplish the first kind of collection processing. There are several specialized reduction functions, which include `min()`, `max()`, `len()`, and `sum()`. There's a general-purpose reduction function as well, `functools.reduce()`.

We'll also consider a `collections.Counter()` function as a kind of reduction operator. It doesn't produce a single scalar value per se, but it does create a new organization of the data that eliminates some of the original structure. At heart, it's a kind of count-group-by operation that has more in common with a counting reduction than with a mapping.

In this chapter, we'll look at reduction functions in more detail. From a purely functional perspective, a reduction is defined recursively. For this reason, we'll look at recursion first before we look at reduction algorithms.

Generally, a functional programming language compiler will optimize a recursive function to transform a call in the tail of the function to a loop. This will dramatically improve performance. From a Python perspective, pure recursion is limited, so we must do the tail-call optimization manually. The tail-call optimization technique available in Python is to use an explicit `for` loop.

We'll look at a number of reduction algorithms including `sum()`, `count()`, `max()`, and `min()`. We'll also look at the `collections.Counter()` function and related `groupby()` reductions. We'll also look at how parsing (and lexical scanning) are proper reductions since they transform sequences of tokens (or sequences of characters) into higher-order collections with more complex properties.

Simple numerical recursions

We can consider all numeric operations to be defined by recursions. For more details, read about the **Peano axioms** that define the essential features of numbers at: `http://en.wikipedia.org/wiki/Peano_axioms`.

From these axioms, we can see that addition is defined recursively using more primitive notions of the next number, or successor of a number, *n*, $S(n)$.

To simplify the presentation, we'll assume that we can define a predecessor function, $P(n)$, such that $n = S(P(n)) = P(S(n))$, as long as $n \neq 0$. This formalizes the idea that a number is the successor of the number's predecessor.

Addition between two natural numbers could be defined recursively as follows:

$$\text{add}(a, b) = \begin{cases} b & \textbf{if } a = 0 \\ \text{add}(P(s), S(b)) & \textbf{if } a \neq 0 \end{cases}$$

If we use the more common $n + 1$ and $n - 1$ instead of $S(n)$ and $P(n)$, we can see that $\text{add}(a, b) = \text{add}(a - 1, b + 1)$.

This translates neatly into Python, as shown in the following command snippet:

```
def add(a: int, b: int) -> int:
    if a == 0:
        return b
    else:
        return add(a-1, b+1)
```

We've rearranged common mathematical notation into Python.

There's no good reason to provide our own functions in Python to do simple addition. We rely on Python's underlying implementation to properly handle arithmetic of various kinds. Our point here is that fundamental scalar arithmetic can be defined recursively, and the definition is very simple to implement.

All of these recursive definitions include at least two cases: the nonrecursive (or *base*) cases where the value of the function is defined directly, and recursive cases where the value of the function is computed from a recursive evaluation of the function with different values.

In order to be sure the recursion will terminate, it's important to see how the recursive case computes values that approach the defined nonrecursive case. There are often constraints on the argument values that we've omitted from the functions here. The `add()` function in the preceding command snippet, for example, can include `assert a>=0 and b>=0` to establish the constraints on the input values.

Without these constraints, starting with `a` set to `-1` can't be guaranteed to approach the nonrecursive case of `a == 0`.

Implementing tail-call optimization

For some functions, the recursive definition is the most succinct and expressive. A common example is the `factorial()` function.

We can see how this is rewritten as a simple recursive function in Python from the following formula:

$$n! = \begin{cases} 1 & \textbf{if } n = 0 \\ n \times (n-1)! & \textbf{if } n > 0 \end{cases}$$

The preceding formula can be executed in Python by using the following commands:

```
def fact(n: int) -> int:
    if n == 0: return 1
    else: return n*fact(n-1)
```

This has the advantage of simplicity. The recursion limits in Python artificially constrain us; we can't do anything larger than about `fact(997)`. The value of 1000! has 2,568 digits and generally exceeds our floating-point capacity; on some systems the floating-point limit is near 10^{300}. Pragmatically, it's common to switch to a `log gamma` function, which works well with large floating-point values.

This function demonstrates a typical tail recursion. The last expression in the function is a call to the function with a new argument value. An optimizing compiler can replace the function call stack management with a loop that executes very quickly.

Since Python doesn't have an optimizing compiler, we're obliged to look at scalar recursions with an eye toward optimizing them. In this case, the function involves an incremental change from *n* to *n*-1. This means that we're generating a sequence of numbers and then doing a reduction to compute their product.

Stepping outside purely functional processing, we can define an imperative facti() calculation as follows:

```
def facti(n: int) -> int:
    if n == 0: return 1
    f = 1
    for i in range(2, n):
        f = f*i
    return f
```

This version of the factorial function will compute values beyond 1000! (2000!, for example, has 5733 digits). It isn't purely functional. We've optimized the tail recursion into a stateful loop depending on the i variable to maintain the state of the computation.

In general, we're obliged to do this in Python because Python can't automatically do the tail-call optimization. There are situations, however, where this kind of optimization isn't actually helpful. We'll look at a few situations.

Leaving recursion in place

In some cases, the recursive definition is actually optimal. Some recursions involve a divide and conquer strategy that minimizes the work. One example of this is the exponentiation by the squaring algorithm. We can state it formally as follows:

$$a^n = \begin{cases} 1 & \textbf{if } n = 0 \\ a \times a^{n-1} & \textbf{if } n \textbf{ is odd} \\ (a^{\frac{n}{2}})^2 & \textbf{if } n \textbf{ is even} \end{cases}$$

We've broken the process into three cases, easily written in Python as a recursion. Look at the following command snippet:

```
def fastexp(a: float, n: int) -> float:
    if n == 0:
        return 1
    elif n % 2 == 1:
        return a*fastexp(a, n-1)
    else:
        t= fastexp(a, n//2)
        return t*t
```

This function has three cases. The base case, the `fastexp(a, 0)` method is defined as having a value of `1`. The other two cases take two different approaches. For odd numbers, the `fastexp()` method is defined recursively. The exponent *n* is reduced by `1`. A simple tail-recursion optimization would work for this case.

For even numbers, however, the `fastexp()` recursion uses `n/2`, chopping the problem into half of its original size. Since the problem size is reduced by a factor of `2`, this case results in a significant speed-up of the processing.

We can't trivially reframe this kind of function into a tail-call optimization loop. Since it's already optimal, we don't really need to optimize it further. The recursion limit in Python would impose the constraint of $n \leq 2^{1000}$, a generous upper bound.

The type hints suggest that this function is designed for float values. It will also work for integer values. Because of Python's type coercion rules, it's simpler to use `float` for a type hint on a function that works with either of the common numeric types.

Handling difficult tail-call optimization

We can look at the definition of **Fibonacci** numbers recursively. The following is one widely used definition for the *nth* Fibonacci number, F_n:

$$F_n = \begin{cases} 0 & \textbf{if } n = 0 \\ 1 & \textbf{if } n = 1 \\ F_{n-1} + F_{n-2} & \textbf{if } n \geq 2 \end{cases}$$

A given Fibonacci number, F_n, is defined as the sum of the previous two numbers, $F_{n-1} + F_{n-2}$. This is an example of multiple recursion: it can't be trivially optimized as a simple tail-recursion. However, if we don't optimize it to a tail-recursion, we'll find it to be too slow to be useful.

The following is a naïve implementation:

```
def fib(n: int) -> int:
    if n == 0: return 0
    if n == 1: return 1
    return fib(n-1) + fib(n-2)
```

This suffers from the multiple recursion problem. When computing the `fib(n)` method, we must compute the `fib(n-1)` and `fib(n-2)` methods. The computation of the `fib(n-1)` method involves a duplicate calculation of the `fib(n-2)` method. The two recursive uses of the Fibonacci function will duplicate the amount of computation being done.

Because of the left-to-right Python evaluation rules, we can evaluate values up to about `fib(1000)`. However, we have to be patient. Very patient.

The following is an alternative, which restates the entire algorithm to use stateful variables instead of a simple recursion:

```
def fibi(n: int) -> int:
    if n == 0: return 0
    if n == 1: return 1
    f_n2, f_n1 = 1, 1
    for _ in range(3, n+1):
        f_n2, f_n1 = f_n1, f_n2+f_n1
    return f_n1
```

Our stateful version of this function counts up from `0`, unlike the recursion, which counts down from the initial value of *n*. This version is considerably faster than the recursive version.

What's important here is that we couldn't trivially optimize the `fib()` function recursion with an obvious rewrite. In order to replace the recursion with an imperative version, we had to look closely at the algorithm to determine how many stateful intermediate variables were required.

Processing collections through recursion

When working with a collection, we can also define the processing recursively. We can, for example, define the `map()` function recursively. The formalism could be stated as follows:

$$\text{map}(f, C) = \begin{cases} [\,] & \textbf{if } len(c) = 0 \\ \text{map}(f, C[:-1]) \text{ append } f(C[-1]) & \textbf{if } len(c) \neq 0 \end{cases}$$

We've defined the mapping of a function to an empty collection as an empty sequence. We've also specified that applying a function to a collection can be defined recursively with a three step expression. First, apply the function to all of the collection except the last element, creating a sequence object. Then apply the function to the last element. Finally, append the last calculation to the previously built sequence.

Following is a purely recursive function version of the older map() function:

```
from typing import Callable, Sequence, Any
def mapr(
        f: Callable[[Any], Any],
        collection: Sequence[Any]
    ) -> List[Any]:
    if len(collection) == 0: return []
    return mapr(f, collection[:-1]) + [f(collection[-1])]
```

The value of the mapr(f, []) method is defined to be an empty list object. The value of the mapr() function with a non-empty list will apply the function to the last element in the list and append this to the list built recursively from the mapr() function applied to the head of the list.

We have to emphasize that this mapr() function actually creates a list object, similar to the older map() function in Python 2. The Python 3 map() function is an iterable; it doesn't create a list object.

While this is an elegant formalism, it still lacks the tail-call optimization required. The tail-call optimization allows us to exceed the recursion depth of 1000 and also performs much more quickly than this naïve recursion.

The use of Callable[[Any], Any] is a weak type hint. To be more clear, it can help to define a domain type variable and a range type variable. We'll include this in the the optimized example

Tail-call optimization for collections

We have two general ways to handle collections: we can use a higher-order function that returns a generator expression, or we can create a function that uses a for loop to process each item in a collection. The two essential patterns are very similar.

Following is a higher-order function that behaves like the built-in map() function:

```
from typing import Callable, Iterable, Iterator, Any, TypeVar
D_ = TypeVar("D_")
R_ = TypeVar("R_")
def mapf(
        f: Callable[[D_], R_],
        C: Iterable[D_]
    ) -> Iterator[R_]:
    return (f(x) for x in C)
```

We've returned a generator expression that produces the required mapping. This uses the explicit `for` in the generator expression as a kind of tail-call optimization.

The source of data `C` has a type hint of `Iterable[D_]` to emphasize that some type will form the domain for the mapping. The transformation function has a hint of `Callable[[D_], R_]` to make it clear that it transforms from some domain type to a range type. The function `float()`, for example, can transform values from the string domain to the float range. The result has the hint of `Iterator[R_]` to show that it iterates over the range type; the result type of the callable function.

Following is a generator function with the same signature and result:

```
def mapg(
        f: Callable[[D_], R_],
        C: Iterable[D_]
    ) -> Iterator[R_]:
    for x in C:
        yield f(x)
```

This uses a complete `for` statement for the tail-call optimization. The results are identical. This version is slightly slower because this involves multiple statements.

In both cases, the result is an iterator over the results. We must do something to materialize a sequence object from an iterable source. For example, here is the `list()` function being used to create a sequence from the iterator:

```
>>> list(mapg(lambda x:2**x, [0, 1, 2, 3, 4]))
[1, 2, 4, 8, 16]
```

For performance and scalability, this kind of tail-call optimization is required in Python programs. It makes the code less than purely functional. However, the benefit far outweighs the lack of purity. In order to reap the benefits of succinct and expressive functional design, it is helpful to treat these less-than-pure functions as if they were proper recursions.

What this means, pragmatically, is that we must avoid cluttering up a collection processing function with additional stateful processing. The central tenets of functional programming are still valid even if some elements of our programs are less than purely functional.

Reductions and folding a collection from many items to one

We can consider the `sum()` function to have the following kind of definition:

We could say that the sum of a collection is 0 for an empty collection. For a non-empty collection the sum is the first element plus the sum of the remaining elements.

$$\text{sum}(C) = \begin{cases} 0 & \textbf{if} \; \text{len}(C) = 0 \\ C[0] + \text{sum}(C[1:]) & \textbf{if} \; \text{len}(C) > 0 \end{cases}$$

Similarly, we can compute the product of a collection of numbers recursively using two cases:

$$\text{prod}(C) = \begin{cases} 1 & \textbf{if} \; \text{len}(C) = 0 \\ C[0] \times \text{prod}(C[1:]) & \textbf{if} \; \text{len}(C) > 0 \end{cases}$$

The base case defines the product of an empty sequence as 1. The recursive case defines the product as the first item times the product of the remaining items.

We've effectively folded in × or + operators between each item of the sequence. Furthermore, we've grouped the items so that processing will be done right to left. This could be called a fold-right way of reducing a collection to a single value.

In Python, the product function can be defined recursively as follows:

```
def prodrc(collection: Sequence[float]) -> float:
    if len(collection) == 0: return 1
    return collection[0] * prodrc(collection[1:])
```

This is technically correct. It's a trivial rewrite from a mathematical notation to Python. However, it is less than optimal because it tends to create a large number of intermediate `list` objects. It's also limited to only working with explicit collections; it can't work easily with `iterable` objects.

Also note that we've used `float` as the generic numeric type hint. This will work for integers and will produce an integer result. It's simpler to use `float` as a generic type hint for numeric functions like this.

We can revise this slightly to work with an iterable, which avoids creating any intermediate `collection` objects. The following is a properly recursive product function that works with an iterable source of data:

```
def prodri(items: Iterator[float]) -> float:
    try:
        head= next(iterable)
    except StopIteration:
        return 1
    return head*prodri(iterable)
```

We can't interrogate an iterable with the `len()` function to see how many elements it has. All we can do is attempt to extract the head of the `iterable` sequence. If there are no items in the sequence, then any attempt to get the head will raise the `StopIteration` exception. If there is an item, then we can multiply this item by the product of the remaining items in the sequence. For a demo, we must explicitly create an iterable from a materialized `sequence` object, using the `iter()` function. In other contexts, we might have an iterable result that we can use. Following is an example:

```
>>> prodri(iter([1,2,3,4,5,6,7]))
5040
```

This recursive definition does not rely on explicit state or other imperative features of Python. While it's more purely functional, it is still limited to working with collections of under 1000 items. Pragmatically, we can use the following kind of imperative structure for reduction functions:

```
def prodi(items: Iterable[float]) -> float:
    p = 1
    for n in iterable:
        p *= n
    return p
```

This avoids any recursion limits. It includes the required tail-call optimization. Furthermore, this will work equally well with either a `Sequence` object or an iterable.

In other functional languages, this is called a `foldl` operation: the operators are folded into the iterable collection of values from left to right. This is unlike the recursive formulations, which are generally called `foldr` operations because the evaluations are done from right to left in the collection.

For languages with optimizing compilers and lazy evaluation, the fold-left and fold-right distinction determines how intermediate results are created. This may have profound performance implications, but the distinction might not be obvious. A fold-left, for example, could immediately consume and process the first elements in a sequence. A fold-right, however, might consume the head of the sequence, but not do any processing until the entire sequence was consumed.

Group-by reduction from many items to fewer

A very common operation is a reduction that groups values by some key or indicator. In **SQL**, this is often called the `SELECT GROUP BY` operation. The raw data is grouped by some column's value and reductions (sometimes aggregate functions) are applied to other columns. The SQL aggregate functions include `SUM`, `COUNT`, `MAX`, and `MIN`.

The statistical summary, called the mode, is a count that's grouped by some independent variable. Python offers us several ways to group data before computing a reduction of the grouped values. We'll start by looking at two ways to get simple counts of grouped data. Then we'll look at ways to compute different summaries of grouped data.

We'll use the trip data that we computed in `Chapter 4`, *Working with Collections*. This data started as a sequence of latitude-longitude waypoints. We restructured it to create legs represented by three tuples of start, end, and distance for the `leg`. The data looks as follows:

```
(((37.5490162, -76.330295), (37.840832, -76.273834), 17.7246),
 ((37.840832, -76.273834), (38.331501, -76.459503), 30.7382),
 ((38.331501, -76.459503), (38.845501, -76.537331), 31.0756), ...
 ((38.330166, -76.458504), (38.976334, -76.473503), 38.8019))
```

A common operation, which can be approached either as a stateful map or as a materialized, sorted object, is computing the mode of a set of data values. When we look at our trip data, the variables are all continuous. To compute a mode, we need to quantize the distances covered. This is also called **binning**: we group the data into different bins. Binning is common in data visualization applications, also. In this case, we'll use 5 nautical miles as the size of each bin.

The quantized distances can be produced with a generator expression:

```
quantized = (5*(dist//5) for start, stop, dist in trip)
```

This will divide each distance by 5—discarding any fractions—and then multiply by 5 to compute a number that represents the distance rounded down to the nearest 5 nautical miles.

Building a mapping with Counter

A mapping like the `collections.Counter` method is a great optimization for doing reductions that create counts (or totals) grouped by some value in the collection. A more typical functional programming solution to grouping data is to sort the original collection, and then use a recursive loop to identify when each group begins. This involves materializing the raw data, performing a $\mathbf{O}(n \log n)$ sort, and then doing a reduction to get the sums or counts for each key.

We'll use the following generator to create a simple sequence of distances transformed into bins:

```
quantized = (5*(dist//5) for start, stop, dist in trip)
```

We divided each distance by 5 using truncated integer division, and then multiplied by 5 to create a value that's rounded down to the nearest 5 miles.

The following expression creates a `mapping` from distance to frequency:

```
from collections import Counter
Counter(quantized)
```

This is a stateful object, that was created by—technically imperative object-oriented programming. Since it looks like a function, however, it seems a good fit for a design based on functional programming ideas.

If we print `Counter(quantized).most_common()` function, we'll see the following results:

```
[(30.0, 15), (15.0, 9), (35.0, 5), (5.0, 5), (10.0, 5), (20.0, 5),
 (25.0, 5), (0.0, 4), (40.0, 3), (45.0, 3), (50.0, 3), (60.0, 3),
 (70.0, 2), (65.0, 1), (80.0, 1), (115.0, 1), (85.0, 1), (55.0, 1),
 (125.0, 1)]
```

The most common distance was about 30 nautical miles. The shortest recorded `leg` was four instances of 0. The longest leg was 125 nautical miles.

Note that your output may vary slightly from this. The results of the `most_common()` function are in order of frequency; equal-frequency bins may be in any order. These five lengths may not always be in the order shown:

```
(35.0, 5), (5.0, 5), (10.0, 5), (20.0, 5), (25.0, 5)
```

Building a mapping by sorting

If we want to implement this without using the `Counter` class, we can use a more functional approach of sorting and grouping. Following is a common algorithm:

```
from typing import Dict, Any, Iterable, Tuple, List, TypeVar
Leg = Tuple[Any, Any, float]
T_ = TypeVar("T_")

def group_sort1(trip: Iterable[Leg]) -> Dict[int, int]:
    def group(
            data: Iterable[T_]
        ) -> Iterable[Tuple[T_, int]]:
        previous, count = None, 0
        for d in sorted(data):
            if d == previous:
                count += 1
            elif previous is not None:  # and d != previous
                yield previous, count
                previous, count = d, 1
            elif previous is None:
                previous, count = d, 1
            else:
                raise Exception("Bad bad design problem.")
        yield previous, count
    quantized = (int(5*(dist//5)) for beg, end, dist in trip)
    return dict(group(quantized))
```

The internal `group()` function steps through the sorted sequence of legs. If a given item has already been seen—it matches the value in `previous`—then the counter is incremented. If a given item does not match the previous value and the previous value is not `None`, then there's been a change in value: emit the previous value and the count, and begin a new accumulation of counts for the new value. The third condition only applies once: if the previous value has never been set, then this is the first value, and we should save it.

The definition of group() provides two important type hints. The source data is an iterable over some type, shown with the type variable T_. In this specific case, it's pretty clear that the value of T_ will be an int; however, the algorithm will work for any Python type. The resulting iterable from the group() function will preserve the type of the source data, and this is made explicit by using the same type variable, T_.

The final line of the group_sort1() function creates a dictionary from the grouped items. This dictionary will be similar to a Counter dictionary. The primary difference is that a Counter() function will have a most_common() method function, which a default dictionary lacks.

The elif previous is None method case is an irksome overhead. Getting rid of this elif clause (and seeing a slight performance improvement) isn't terribly difficult.

To remove the extra elif clause, we need to use a slightly more elaborate initialization in the internal group() function:

```
def group(data: Iterable[T_]) -> Iterable[Tuple[T_, int]]:
    sorted_data = iter(sorted(data))
    previous, count = next(sorted_data), 1
    for d in sorted_data:
        if d == previous:
            count += 1
        elif previous is not None: # and d != previous
            yield previous, count
            previous, count = d, 1
        else:
            raise Exception("Bad bad design problem.")
    yield previous, count
```

This picks the first item out of the set of data to initialize the previous variable. The remaining items are then processed through the loop. This design shows a loose parallel with recursive designs, where we initialize the recursion with the first item, and each recursive call provides either a next item or None to indicate that no items are left to process.

We can also do this with itertools.groupby(). We'll look at this function closely in Chapter 8, *The Itertools Module*.

Grouping or partitioning data by key values

There are no limits to the kinds of reductions we might want to apply to grouped data. We might have data with a number of independent and dependent variables. We can consider partitioning the data by an independent variable and computing summaries such as maximum, minimum, average, and standard deviation of the values in each partition.

The essential trick to doing more sophisticated reductions is to collect all of the data values into each group. The `Counter()` function merely collects counts of identical items. We want to create sequences of the original items based on a key value.

Looked at it in a more general way, each five-mile bin will contain the entire collection of legs of that distance, not merely a count of the legs. We can consider the partitioning as a recursion or as a stateful application of `defaultdict(list)` objects. We'll look at the recursive definition of a `groupby()` function, since it's easy to design.

Clearly, the `groupby(C, key)` method for an empty collection, `C`, is the empty dictionary, `dict()`. Or, more usefully, the empty `defaultdict(list)` object.

For a non-empty collection, we need to work with item `C[0]`, the head, and recursively process sequence `C[1:]`, the tail. We can use the `head, *tail = C` command to do this parsing of the collection, as follows:

```
>>> C= [1,2,3,4,5]
>>> head, *tail= C
>>> head
1
>>> tail
[2, 3, 4, 5]
```

We need to do the `dict[key(head)].append(head)` method to include the head element in the resulting dictionary. And then we need to do the `groupby(tail,key)` method to process the remaining elements.

We can create a function as follows:

```
from typing import Callable, Sequence, Dict, List, TypeVar
S_ = TypeVar("S_")
K_ = TypeVar("K_")
def group_by(
        key: Callable[[S_], K_],
        data: Sequence[S_]
    ) -> Dict[K_, List[S_]]:

    def group_into(
```

```
        key: Callable[[S_], K_],
        collection: Sequence[S_],
        dictionary: Dict[K_, List[S_]]
    ) -> Dict[K_, List[S_]]:
    if len(collection) == 0:
        return dictionary
    head, *tail = collection
    dictionary[key(head)].append(head)
    return group_into(key, tail, dictionary)

return group_into(key, data, defaultdict(list))
```

The interior function `group_into()` handles the essential recursive definition. An empty value for `collection` returns the provided dictionary, untouched. A non-empty collection is partitioned into a head and tail. The head is used to update the dictionary. The tail is then used, recursively, to update the dictionary with all remaining elements.

The type hints make an explicit distinction between the type of the source objects `S_` and the type of the key `K_`. The function provided as the `key` parameter must be a callable that returns a value of the key type `K_`, given an object of the source type `S_`. In many of the examples, a function to extract the distance from a `Leg` object will be be shown. This is a `Callable[[S_], K_]` where the source type `S_` is the `Leg` object and the key type `K_` is the `float` value.

We can't easily use Python's default values to collapse this into a single function. We explicitly cannot use the following command snippet:

```
def group_by(key, data, dictionary=defaultdict(list)):
```

If we try this, all uses of the `group_by()` function share one common `defaultdict(list)` object. This doesn't work because Python builds the default value just once. Mutable objects as default values rarely do what we want. Rather than try to include more sophisticated decision-making to handle an immutable default value (for example, `None`), we prefer to use a nested function definition. The `wrapper()` function properly initializes the arguments to the interior function.

We can group the data by distance as follows:

```
binned_distance = lambda leg: 5*(leg[2]//5)
by_distance = group_by(binned_distance, trip)
```

We've defined a simple, reusable `lambda` that puts our distances into 5 nautical mile bins. We then grouped the data using the provided `lambda`.

We can examine the binned data as follows:

```
import pprint
for distance in sorted(by_distance):
    print(distance)
    pprint.pprint(by_distance[distance])
```

The following is what the output looks like:

```
0.0
[((35.505665, -76.653664), (35.508335, -76.654999), 0.1731),
 ((35.028175, -76.682495), (35.031334, -76.682663), 0.1898),
 ((25.4095, -77.910164), (25.425833, -77.832664), 4.3155),
 ((25.0765, -77.308167), (25.080334, -77.334), 1.4235)]
5.0
[((38.845501, -76.537331), (38.992832, -76.451332), 9.7151),
 ((34.972332, -76.585167), (35.028175, -76.682495), 5.8441),
 ((30.717167, -81.552498), (30.766333, -81.471832), 5.103),
 ((25.471333, -78.408165), (25.504833, -78.232834), 9.7128),
 ((23.9555, -76.31633), (24.099667, -76.401833), 9.844)] ...
125.0
[((27.154167, -80.195663), (29.195168, -81.002998), 129.7748)]
```

The `partition()` function can be written as an iteration as follows:

```
from typing import Callable, Dict, List, TypeVar
S_ = TypeVar("S_")
K_ = TypeVar("K_")
def partition(
        key: Callable[[S_], K_],
        data: Iterable[S_]
    ) -> Dict[K_, List[S_]]:
    dictionary: Dict[K_, List[S_]] = defaultdict(list)
    for head in data:
        dictionary[key(head)].append(head)
    return dictionary
```

When doing the tail-call optimization, the essential line of the code in the imperative version will match the recursive definition. We've highlighted that line to emphasize that the rewrite is intended to have the same outcome. The rest of the structure represents the tail-call optimization we've adopted as a common way to work around the Python limitations.

The type hints emphasize the distinction between the source type `S_` and the key type `K_`. Note that the result of `defaultdict(list)` requires an additional type hint of `Dict[K_, List[S_]]` to help the **mypy** tool confirm that this code works. Without the hint, this will produce an `error: Need type annotation for variable` message. A `defaultdict` can have almost any combination of types; without the hint, it's impossible to be sure the variable is being used properly.

This hint can also be provided with a comment as follows:

```
dictionary = defaultdict(list)  # type: Dict[K_, List[S_]]
```

This was required for an older version of the **pylint** tool. Versions after 1.8 are recommended.

Writing more general group-by reductions

Once we have partitioned the raw data, we can compute various kinds of reductions on the data elements in each partition. We might, for example, want the northernmost point for the start of each leg in the distance bins.

We'll introduce some helper functions to decompose the tuple as follows:

```
start = lambda s, e, d: s
end = lambda s, e, d: e
dist = lambda s, e, d: d
latitude = lambda lat, lon: lat
longitude = lambda lat, lon: lon
```

Each of these helper functions expects a `tuple` object to be provided using the `*` operator to map each element of the tuple to a separate parameter of the `lambda`. Once the tuple is expanded into the `s`, `e`, and `p` parameters, it's reasonably obvious to return the proper parameter by name. It's much clearer than trying to interpret the `tuple_arg[2]` method.

The following is how we use these helper functions:

```
>>> point = ((35.505665, -76.653664), (35.508335, -76.654999),
 0.1731)
>>> start(*point)
(35.505665, -76.653664)
>>> end(*point)
(35.508335, -76.654999)
>>> dist(*point)
0.1731
>>> latitude(*start(*point))
```

```
35.505665
```

Our initial point object is a nested three tuple with `(0)`—a starting position, `(1)`—the ending position, and `(2)`—the distance. We extracted various fields using our helper functions.

Given these helpers, we can locate the northernmost starting position for the legs in each bin:

```
for distance in sorted(by_distance):
    print(
        distance,
        max(by_distance[distance],
            key=lambda pt: latitude(*start(*pt)))
    )
```

The data that we grouped by distance included each leg of the given distance. We supplied all of the legs in each bin to the `max()` function. The `key` function we provided to the `max()` function extracted just the latitude of the starting point of the leg.

This gives us a short list of the northernmost legs of each distance, as follows:

```
0.0 ((35.505665, -76.653664), (35.508335, -76.654999), 0.1731)
5.0 ((38.845501, -76.537331), (38.992832, -76.451332), 9.7151)
10.0 ((36.444168, -76.3265), (36.297501, -76.217834), 10.2537)
...
125.0 ((27.154167, -80.195663), (29.195168, -81.002998), 129.7748)
```

Writing higher-order reductions

We'll look at an example of a higher-order reduction algorithm here. This will introduce a rather complex topic. The simplest kind of reduction develops a single value from a collection of values. Python has a number of built-in reductions, including `any()`, `all()`, `max()`, `min()`, `sum()`, and `len()`.

As we noted in `Chapter 4`, *Working with Collections*, we can do a great deal of statistical calculation if we start with a few simple reductions such as the following:

```
def s0(data: Sequence) -> float:
    return sum(1 for x in data)  # or len(data)
def s1(data: Sequence) -> float:
    return sum(x for x in data)  # or sum(data)
def s2(data: Sequence) -> float:
    return sum(x*x for x in data)
```

This allows us to define mean, standard deviation, normalized values, correction, and even least-squares linear regression, using a few simple functions.

The last of our simple reductions, `s2()`, shows how we can apply existing reductions to create higher-order functions. We might change our approach to be more like the following:

```
from typing import Callable, Iterable, Any
def sum_f(
        function: Callable[[Any], float],
        data: Iterable) -> float:
    return sum(function(x) for x in data)
```

We've added a function that we'll use to transform the data. This function computes the sum of the transformed values.

Now we can apply this function in three different ways to compute the three essential sums as follows:

```
N = sum_f(lambda x: 1, data)  # x**0
S = sum_f(lambda x: x, data)  # x**1
S2 = sum_f(lambda x: x*x, data)  # x**2
```

We've plugged in a small `lambda` to compute $\sum_{x \in X} x^0 = \sum_{x \in X} 1$, which is the count, $\sum_{x \in X} x^1 = \sum_{x \in X} x$, the sum, and $\sum_{x \in X} x^2$, the sum of the squares, which we can use to compute standard deviation.

A common extension to this includes a filter to reject raw data that is unknown or unsuitable in some way. We might use the following command to reject bad data:

```
def sum_filter_f(
        filter_f: Callable,
        function: Callable, data: Iterable) -> Iterator:
    return sum(function(x) for x in data if filter_f(x))
```

Execution of the following command snippet allows us to do things such as reject `None` values in a simple way:

```
count_ = lambda x: 1
sum_ = lambda x: x
valid = lambda x: x is not None
N = sum_filter_f(valid, count_, data)
```

This shows how we can provide two distinct `lambda` to our `sum_filter_f()` function. The `filter` argument is a `lambda` that rejects `None` values, we've called it `valid` to emphasize its meaning. The `function` argument is a `lambda` that implements a `count` or a `sum` method. We can easily add a `lambda` to compute a sum of squares.

It's important to note that this function is similar to other examples in that it actually returns a function rather than a value. This is one of the defining characteristics of higher-order functions, and is pleasantly simple to implement in Python.

Writing file parsers

We can often consider a file parser to be a kind of reduction. Many languages have two levels of definition: the lower-level tokens in the language and the higher-level structures built from those tokens. When looking at an XML file, the tags, tag names, and attribute names form this lower-level syntax; the structures which are described by XML form a higher-level syntax.

The lower-level lexical scanning is a kind of reduction that takes individual characters and groups them into tokens. This fits well with Python's generator function design pattern. We can often write functions that look as follows:

```
def lexical_scan(some_source):
    for char in some_source:
        if some pattern completed: yield token
        else: accumulate token
```

For our purposes, we'll rely on lower-level file parsers to handle this for us. We'll use the CSV, JSON, and XML packages to manage these details. We'll write higher-level parsers based on these packages.

We'll still rely on a two-level design pattern. A lower-level parser will produce a useful canonical representation of the raw data. It will be an iterator over tuples of text. This is compatible with many kinds of data files. The higher-level parser will produce objects useful for our specific application. These might be tuples of numbers, or namedtuples, or perhaps some other class of immutable Python objects.

We provided one example of a lower-level parser in `Chapter 4`, *Working with Collections*. The input was a KML file; KML is an XML representation of geographic information. The essential features of the parser look similar to the following command snippet:

```
def comma_split(text: str) -> List[str]:
    return text.split(",")
    def row_iter_kml(file_obj: TextIO) -> Iterator[List[str]]:
```

```
    ns_map = {
        "ns0": "http://www.opengis.net/kml/2.2",
        "ns1": "http://www.google.com/kml/ext/2.2"}
    xpath = (
        "./ns0:Document/ns0:Folder/"
        "ns0:Placemark/ns0:Point/ns0:coordinates")
    doc = XML.parse(file_obj)
    return (
        comma_split(cast(str, coordinates.text))
        for coordinates in doc.findall(xpath, ns_map)
    )
```

The bulk of the `row_iter_kml()` function is the XML parsing that allows us to use the `doc.findall()` function to iterate through the `<ns0:coordinates>` tags in the document. We've used a function named `comma_split()` to parse the text of this tag into a three tuple of values.

The `cast()` function is only present to provide evidence to **mypy** that the value of `coordinates.text` is a `str` object. The default definition of the text attribute is `Union[str, bytes]`; in this application, the data will be `str` exclusively. The `cast()` function doesn't do any run-time processing; it's a type hint that's used by **mypy**.

This function focused on working with the normalized XML structure. The document is close to the database designer's definitions of First Normal Form: each attribute is atomic (a single value), and each row in the XML data has the same columns with data of a consistent type. The data values aren't fully atomic, however: we have to split the points on the `,` to separate longitude, latitude, and altitude into atomic string values. However, the text is completely consistent, making it a close fit with first normal form.

A large volume of data—XML tags, attributes, and other punctuation—is reduced to a somewhat smaller volume, including just floating-point latitude and longitude values. For this reason, we can think of parsers as a kind of reduction.

We'll need a higher-level set of conversions to map the tuples of text into floating-point numbers. Also, we'd like to discard altitude, and reorder longitude and latitude. This will produce the application-specific tuple we need. We can use functions as follows for this conversion:

```
def pick_lat_lon(
        lon: Any, lat: Any, alt: Any) -> Tuple[Any, Any]:
    return lat, lon

def float_lat_lon(
        row_iter: Iterator[Tuple[str, ...]]
    ) -> Iterator[Tuple[float, ...]]:
```

```
    return (
        tuple(
            map(float, pick_lat_lon(*row))
        )
        for row in row_iter
    )
```

The essential tool is the `float_lat_lon()` function. This is a higher-order function that returns a generator expression. The generator uses the `map()` function to apply the `float()` function conversion to the results of `pick_lat_lon()` function, and the `*row` argument to assign each member of the row `tuple` to a different parameter of the `pick_lat_lon()` function. This only works when each row is a three tuple. The `pick_lat_lon()` function then returns a tuple of the selected items in the required order.

We can use this parser as follows:

```
name = "file:./Winter%202012-2013.kml"
with urllib.request.urlopen(name) as source:
    trip = tuple(float_lat_lon(row_iter_kml(source)))
```

This will build a tuple-of-tuple representation of each waypoint along the path in the original KML file. It uses a low-level parser to extract rows of text data from the original representation. It uses a high-level parser to transform the text items into more useful tuples of floating-point values. In this case, we have not implemented any validation.

Parsing CSV files

In Chapter 3, *Functions, Iterators and Generators*, we saw another example where we parsed a CSV file that was not in a normalized form: we had to discard header rows to make it useful. To do this, we used a simple function that extracted the header and returned an iterator over the remaining rows.

The data looks as follows:

```
Anscombe's quartet
I  II  III  IV
x  y  x  y  x  y  x  y
10.0  8.04  10.0  9.14  10.0  7.46  8.0  6.58
8.0  6.95  8.0  8.14  8.0  6.77  8.0  5.76
...
5.0  5.68  5.0  4.74  5.0  5.73  8.0  6.89
```

The columns are separated by tab characters. Plus there are three rows of headers that we can discard.

Here's another version of that CSV-based parser. We've broken it into three functions. The first, `row_iter()` function, returns the iterator over the rows in a tab-delimited file. The function looks as follows:

```
def row_iter_csv(source: TextIO):
    rdr= csv.reader(source, delimiter="\t")
    return rdr
```

This is a simple wrapper around the CSV parsing process. When we look back at the previous parsers for XML and plain text, this was the kind of thing that was missing from those parsers. Producing an iterable over row tuples can be a common feature of parsers for normalized data.

Once we have a row of tuples, we can pass rows that contain usable data and reject rows that contain other metadata, such as titles and column names. We'll introduce a helper function that we can use to do some of the parsing, plus a `filter()` function to validate a row of data.

Following is the conversion:

```
from typing import Optional, Text
def float_none(data: Optional[Text]) -> Optional[float]:
    try:
        data_f= float(data)
        return data_f
    except ValueError:
        return None
```

This function handles the conversion of a single `string` to `float` values, converting bad data to a `None` value. The type hints of `Optional[Text]` and `Optional[float]` express the ideas of having a value of the given type or having a value of the same type as `None`.

We can embed the `float_none()` function in a mapping so that we convert all columns of a row to a `float` or `None` value. A `lambda` for this looks as follows:

```
from typing import Callable, List, Optional
R_Text = List[Optional[Text]]
R_Float = List[Optional[float]]

float_row: Callable[[R_Text], R_Float] \
    = lambda row: list(map(float_none, row))
```

Two type hints are used to make the definition of `float_row` explicit. The `R_Text` hint defines the text version of a row of data. It will be a list that mixes text values with `None` values. The `R_Float` hint defines the floating-point version of a row of data.

Following is a row-level validator based on the use of the `all()` function to assure that all values are `float` (or none of the values are `None`):

```
all_numeric: Callable[[R_Float], bool] \
    = lambda row: all(row) and len(row) == 8
```

This lambda is a kind of reduction, transforming a row of floating-point values to a Boolean value if all values are not "falsy" (that is, neither `None` nor zero) and there are exactly eight values.

The simplistic `all_numeric` function conflates zero and `None`. A more sophisticated test would rely on something such as `not any(item is None for item in row)`. The rewrite is left as an exercise for the reader.

The essential design is to create row-based elements that can be combined to create more complete algorithms for parsing an input file. The foundational functions iterate over tuples of text. These are combined to convert and validate the converted data. For the cases where files are either in first normal form (all rows are the same) or a simple validator can reject the extraneous rows, this design pattern works out nicely.

All parsing problems aren't quite this simple, however. Some files have important data in header or trailer rows that must be preserved, even though it doesn't match the format of the rest of the file. These non-normalized files will require a more sophisticated parser design.

Parsing plain text files with headers

In `Chapter 3`, *Functions, Iterators, and Generators*, the `Crayola.GPL` file was presented without showing the parser. This file looks as follows:

```
GIMP Palette
Name: Crayola
Columns: 16
#
239 222 205  Almond
205 149 117  Antique Brass
```

We can parse a text file using regular expressions. We need to use a filter to read (and parse) header rows. We also want to return an iterable sequence of data rows. This rather complex two-part parsing is based entirely on the two-part—head and tail—file structure.

Following is a low-level parser that handles both the four lines of heading and the long tail:

```
Head_Body = Tuple[Tuple[str, str], Iterator[List[str]]]
def row_iter_gpl(file_obj: TextIO) -> Head_Body:
    header_pat = re.compile(
        r"GIMP Palette\nName:\s*(.*?)\nColumns:\s*(.*?)\n#\n",
        re.M)

    def read_head(
            file_obj: TextIO
        ) -> Tuple[Tuple[str, str], TextIO]:
        match = header_pat.match(
            "".join(file_obj.readline() for _ in range(4))
        )
        return (match.group(1), match.group(2)), file_obj

    def read_tail(
            headers: Tuple[str, str],
            file_obj: TextIO) -> Head_Body:
        return (
            headers,
            (next_line.split() for next_line in file_obj)
        )

    return read_tail(*read_head(file_obj))
```

The `Head_Body` type definition summarizes the overall goal of the row iterator. The result is a two-tuple. The first item is a two-tuple with details from the file header. The second item is a iterator that provides the text items for a color definition. This `Head_Body` type hint is used in two places in this function definition.

The `header_pat` regular expression parses all four lines of the header. There are instances of `()` in the expression to extract the name and column information from the header.

There are two internal functions for parsing different parts of the file. The `read_head()` function parses the header lines and returns interesting text and a `TextIO` object that can be used for the rest of the parsing. It does this by reading four lines and merging them into a single long string. This is then parsed with the `header_pat` regular expression.

The idea of returning the iterator from one function to be used in another function is a pattern for passing an explicitly stateful object from one function to another. It is a minor simplification because all of the arguments for the `read_tail()` function are the results from the `read_head()` function.

The `read_tail()` function parses the iterator over the remaining lines. These lines are merely split on spaces, since that fits the description of the GPL file format.

For more information, visit the following link:

`https://code.google.com/p/grafx2/issues/detail?id=518.`

Once we've transformed each line of the file into a canonical tuple-of-strings format, we can apply the higher level of parsing to this data. This involves conversion and (if necessary) validation.

The following is a higher-level parser command snippet:

```
from typing import NamedTuple
class Color(NamedTuple):
    red: int
    blue: int
    green: int
    name: str

def color_palette(
        headers: Tuple[str, str],
        row_iter: Iterator[List[str]]
    ) -> Tuple[str, str, Tuple[Color, ...]]:
    name, columns = headers
    colors = tuple(
        Color(int(r), int(g), int(b), " ".join(name))
        for r, g, b, *name in row_iter)
    return name, columns, colors
```

This function will work with the output of the lower-level `row_iter_gpl()` parser: it requires the headers and the iterator. This function will use the multiple assignment to separate the `color` numbers and the remaining words into four variables, `r`, `g`, `b`, and `name`. The use of the `*name` parameter ensures that all remaining values will be assigned to the `name` variable as a `tuple`. The `" ".join(name)` method then concatenates the words into a single space-separated string.

the following is how we can use this two-tier parser:

```
with open("crayola.gpl") as source:
    name, cols, colors = color_palette(
        *row_iter_gpl(source)
    )
print(name, cols, colors)
```

We've applied the higher-level parser to the results of the lower-level parser. This will return the headers and a tuple built from the sequence of `Color` objects.

Summary

In this chapter, we've looked at two significant functional programming topics. We've looked at recursions in some detail. Many functional programming language compilers will optimize a recursive function to transform a call in the tail of the function to a loop. In Python, we must do the tail-call optimization manually by using an explicit `for` loop, instead of a purely function recursion.

We've also looked at reduction algorithms, including `sum()`, `count()`, `max()`, and `min()` functions. We looked at the `collections.Counter()` function and related `groupby()` reductions.

We've also looked at how parsing (and lexical scanning) are similar to reductions since they transform sequences of tokens (or sequences of characters) into higher-order collections with more complex properties. We've examined a design pattern that decomposes parsing into a lower level and tries to produce tuples of raw strings, and a higher level that creates more useful application objects.

In the next chapter, we'll look at some techniques appropriate to working with namedtuples and other immutable data structures. We'll look at techniques that make stateful objects unnecessary. While stateful objects aren't purely functional, the idea of a class hierarchy can be used to package related method function definitions.

7
Additional Tuple Techniques

Many of the examples we've looked at have either been functions using atomic (or scalar) objects, or relatively simple structures built from small tuples. We can often exploit Python's immutable `typing.NamedTuple` as a way to build complex data structures.

One of the beneficial features of object-oriented programming is the ability to create complex data structures incrementally. In some respects, an object can be viewed as a cache for results of functions; this will often fit well with functional design patterns. In other cases, the object paradigm provides for property methods that include sophisticated calculations to derive data from an object's properties. This is also a good fit for functional design ideas.

In some cases, however, object class definitions are used statefully to create complex objects. We'll look at a number of alternatives that provide similar features without the complexities of objects with state changes. We can identify stateful class definitions and then include meta-properties for valid or required ordering of method function calls. Statements such as, *if* `X.p()` *is called before* `X.q()`*, the results are undefined* are outside the formalism of the language and are meta-properties of a class. Sometimes, stateful classes include the overhead of explicit assertions and error checking to assure that methods are used in the proper order. If we avoid stateful classes, we eliminate these kinds of overheads.

In this chapter we'll look at the following:

- How we use and how we create `NamedTuple`.
- Ways that immutable `NamedTuple` can be used instead of stateful object classes.
- Some techniques to write generic functions outside any polymorphic class definition. Clearly, we can rely on `Callable` classes to create a polymorphic class hierarchy. In some cases, this might be a needless overhead in a functional design.

Using tuples to collect data

In `Chapter 3`, *Functions, Iterators, and Generators,* we showed two common techniques to work with tuples. We've also hinted at a third way to handle complex structures. We can do any of the following techniques, depending on the circumstances:

- Use lambdas (or functions) to select a named item using the index
- Use lambdas (or functions) with the argument to assign a tuple items to parameter names
- Use named tuples to select an item by attribute name or index

Our trip data, introduced in `Chapter 4`, *Working with Collections,* has a rather complex structure. The data started as an ordinary time series of position reports. To compute the distances covered, we transposed the data into a sequence of legs with a start position, end position, and distance as a nested three-tuple.

Each item in the sequence of legs looks as follows as a three-tuple:

```
first_leg = (
    (37.549016, -76.330295),
    (37.840832, -76.273834),
    17.7246)
```

The first two items are the starting and ending points. The third item is the distance between the points. This is a short trip between two points on the Chesapeake Bay.

A nested tuple of tuples can be rather difficult to read; for example, expressions such as `first_leg[0][0]` aren't very informative.

Let's look at the three alternatives for selected values out of a `tuple`. The first technique involves defining some simple selection functions that can pick items from a `tuple` by index position:

```
start = lambda leg: leg[0]
end = lambda leg: leg[1]
distance = lambda leg: leg[2]
latitude = lambda pt: pt[0]
longitude = lambda pt: pt[1]
```

With these definitions, we can use `latitude(start(first_leg))` to refer to a specific piece of data. It looks like this code example:

```
>>> latitude(start(first_leg))
29.050501
```

These definitions don't provide much guidance on the data types involved. We can use a simple naming convention to make this a bit more clear. The following are some examples of selection functions that use a suffix:

```
start_point = lambda leg: leg[0]
distance_nm = lambda leg: leg[2]
latitude_value = lambda point: point[0]
```

When used judiciously, this can be helpful. It can also degenerate into an elaborately complex Hungarian notation as a prefix (or suffix) of each variable.

It's awkward to provide type hints for lambdas. The following shows how this can be approached:

```
>>> from typing import Tuple, Callable
>>> Point = Tuple[float, float]
>>> Leg = Tuple[Point, Point, float]
>>> start: Callable[[Leg], Point] = lambda leg: leg[0]
```

The type hint is provided as part of the assignment statement. This tells **mypy** that the object named `start` is a callable function that accepts a single parameter of a type named `Leg` and returns a result of the `Point` type.

The second technique uses the `*parameter` notation to conceal some details of the index positions. The following are some selection functions that use the `*` notation:

```
start = lambda start, end, distance: start
end = lambda start, end, distance: end
distance = lambda start, end, distance: distance
latitude = lambda lat, lon: lat
longitude = lambda lat, lon: lon
```

With these definitions, we can use `latitude(*start(*first_leg))` to refer to a specific piece of data. It looks like this code example:

```
>>> latitude(*start(*first_leg))
29.050501
```

This has the advantage of clarity in the function definitions. The association between position and name is given by the list of parameter names. It can look a little odd to see the `*` operator in front of the `tuple` arguments to these selection functions. This operator is useful because it maps each item in a tuple to a parameter of the function.

While these are very functional, the syntax for selecting individual attributes can be confusing. Python offers an object-oriented alternative, the named tuple.

Using named tuples to collect data

The third technique for collecting data into a complex structure is the named tuple. The idea is to create an object that is a tuple as well as a structure with named attributes. There are two variations available:

- The `namedtuple` function in the `collections` module.
- The `NamedTuple` base class in the `typing` module. We'll use this almost exclusively because it allows explicit type hinting.

In the example from the previous section, we have nested namedtuple classes such as the following:

```
from typing import NamedTuple

class Point(NamedTuple):
    latitude: float
    longitude: float

class Leg(NamedTuple):
    start: Point
    end: Point
    distance: float
```

This changes the data structure from simple anonymous tuples to named tuples with type hints provided for each attribute. Here's an example:

```
>>> first_leg = Leg(
... Point(29.050501, -80.651169),
... Point(27.186001, -80.139503),
... 115.1751)
>>> first_leg.start.latitude
29.050501
```

The `first_leg` object was built as the `Leg` subclass of the `NamedTuple` class that contains two other named tuple objects and a float value. Using `first_leg.start.latitude` will fetch a particular piece of data from inside the tuple structure. The change from prefix function names to postfix attribute names can be seen as a helpful emphasis. It can also be seen as a confusing shift in the syntax.

Replacing simple `tuple()` functions with appropriate `Leg()` or `Point()` function calls is important. This changes the processing that builds the data structure. It provides an explicitly named structure with type hints that can be checked by the **mypy** tool.

For example, take a look at the following code snippet to create point pairs from source data:

```
from typing import Tuple, Iterator, List

def float_lat_lon_tuple(
        row_iter: Iterator[List[str]]
    ) -> Iterator[Tuple]:
    return (
        tuple(*map(float, pick_lat_lon(*row)))
        for row in row_iter
    )
```

This replies in an iterator object that produces a list of strings. A CSV reader, or KML reader can do this. The `pick_lat_lon()` function picks two values from the row. The `map()` function applies the `float()` function to the picked values. The result becomes a simple tuple.

The preceding code would be changed to the following code snippet to create `Point` objects:

```
def float_lat_lon(
        row_iter: Iterator[List[str]]
    ) -> Iterator[Point]:
    return (
        Point(*map(float, pick_lat_lon(*row)))
        for row in row_iter
    )
```

The `tuple()` function has been replaced with the `Point()` constructor. The data type that is returned is revised to be `Iterator[Point]`. It's clear that this function builds `Point` objects instead of anonymous tuples of floating-point coordinates.

Similarly, we can introduce the following to build the complete trip of `Leg` objects:

```
from typing import cast, TextIO, Tuple, Iterator, List
from Chapter_6.ch06_ex3 import row_iter_kml
from Chapter_4.ch04_ex1 import legs, haversine

source = "file:./Winter%202012-2013.kml"
def get_trip(url: str=source) -> List[Leg]:
    with urllib.request.urlopen(url) as source:
        path_iter = float_lat_lon(row_iter_kml(
            cast(TextIO, source)
        ))
        pair_iter = legs(path_iter)
```

```
        trip_iter = (
            Leg(start, end, round(haversine(start, end), 4))
            for start, end in pair_iter
        )
        trip = list(trip_iter)
    return trip
```

The processing is a sequence of generator expressions. The `path_iter` object uses two generator functions, `row_iter_kml()` and `float_lat_lon()` to read the rows from a KML file, pick fields, and convert them to `Point` objects. The `pair_iter()` object uses the `legs()` generator function to yield pairs of `Point` objects showing the start and end of each leg.

The `trip_iter` generator expression creates the final `Leg` objects from pairs of `Point` objects. These generated objects are consumed by the `list()` function to create a single list of `Legs`. The `haversine()` function from `Chapter 4`, *Working with Collections*, is used to compute the distance.

The `cast()` function is used here to inform the **mypy** tool that the `source` object is expected to be a `TextIO` instance. The `cast()` function is a type hint; it has no runtime effect. It's required because the `urlopen()` function is defined as a `Union[HTTPResponse, addinfourl]`. An `addinfourl` object is a `BinaryIO`. The `csv.reader()` expects a `List[str]` as an input, it requires text instead of bytes provided by `urlopen()`. For simple CSV files, the distinction between bytes and UTF-8 encoded text is minor, and the `cast()` expedient works.

To work properly with bytes, it's essential to use the `codecs` module to translate the bytes into proper text. The following code can be used:

```
cast(TextIO, codecs.getreader('utf-8')(cast(BinaryIO, source)))
```

The innermost expression is a `cast()` type hint used to make the `source` object appear as a `BinaryIO` type to the **mypy** tool. The `codecs.getreader()` function locates the proper class of readers to handle the `utf-8` encoding. An instance of this class is built using the `source` object to create a reader.

The resulting object is a `StreamReader`. The outermost `cast()` function is a hint for the **mypy** tool to treat the `StreamReader` as a `TextIO` instance. The reader created by `codecs.getreader()` is the essential ingredient in decoding files of bytes into proper text. The other casts are type hints for the **mypy** tool.

The `trip` object is a sequence of `Leg` instances. It will look as follows when we try to print it:

```
(Leg(start=Point(latitude=37.549016, longitude=
 -76.330295), end=Point(latitude=37.840832, longitude=
 -76.273834), distance=17.7246),
 Leg(start=Point(latitude=37.840832, longitude=-76.273834),
 end=Point(latitude=38.331501, longitude=-76.459503),
 distance=30.7382),
 ...
 Leg(start=Point(latitude=38.330166, longitude=-76.458504),
 end=Point(latitude=38.976334, longitude=-76.473503),
 distance=38.8019))
```

It's important to note that the `haversine()` function was written to use simple tuples. We've reused this function with a `NamedTuples` class. As we carefully preserved the order of the arguments, this small change in representation was handled gracefully by Python.

In most cases, the `NamedTuple` function adds clarity. The use of `NamedTuple` will lead to a change from function-like prefix syntax to object-like suffix syntax.

Building named tuples with functional constructors

There are three ways we can build `NamedTuple` instances. The choice of technique we use is generally based on how much additional information is available at the time of object construction.

We've shown two of the three techniques in the examples in the previous section. We'll emphasize the design considerations here. It includes the following choices:

- We can provide the parameter values according to their positions. This works out well when there are one or more expressions that we were evaluating. We used it when applying the `haversine()` function to the `start` and `end` points to create a `Leg` object:

```
Leg(start, end, round(haversine(start, end), 4))
```

- We can use the `*` argument notation to assign parameters according to their positions in a tuple. This works out well when we're getting the arguments from another iterable or an existing tuple. We used it when using `map()` to apply the `float()` function to the `latitude` and `longitude` values:

```
Point(*map(float, pick_lat_lon(*row)))
```

- We can use explicit keyword assignment. While not used in the previous example, we might see something like this as a way to make the relationships more obvious:

```
Point(longitude=float(row[0]), latitude=float(row[1]))
```

It's helpful to have the flexibility of a variety of ways of creating named tuple instances. This allows us to more easily transform the structure of data. We can emphasize features of the data structure that are relevant for reading and understanding the application. Sometimes, the index number of 0 or 1 is an important thing to emphasize. Other times, the order of `start`, `end`, and `distance` is important.

Avoiding stateful classes by using families of tuples

In several previous examples, we've shown the idea of **Wrap-Unwrap** design patterns that allow us to work with anonymous and named tuples. The point of this kind of design is to use immutable objects that wrap other immutable objects instead of mutable instance variables.

A common statistical measure of correlation between two sets of data is the Spearman rank correlation. This compares the rankings of two variables. Rather than trying to compare values, which might have different scales, we'll compare the relative orders. For more information,
visit: `http://en.wikipedia.org/wiki/Spearman%27s_rank_correlation_coefficient`.

Computing the Spearman rank correlation requires assigning a rank value to each observation. It seems like we should be able to use `enumerate(sorted())` to do this. Given two sets of possibly correlated data, we can transform each set into a sequence of rank values and compute a measure of correlation.

We'll apply the Wrap-Unwrap design pattern to do this. We'll `wrap` data items with their rank for the purposes of computing the correlation coefficient.

In `Chapter 3`, *Functions, Iterators, and Generators*, we showed how to parse a simple dataset. We'll extract the four samples from that dataset as follows:

```
>>> from Chapter_3.ch03_ex5 import (
...     series, head_map_filter, row_iter)
>>> with open("Anscombe.txt") as source:
...     data = list(head_map_filter(row_iter(source)))
```

The resulting collection of data has four different series of data combined in each row. The `series()` function will extract the pairs for a given series from the overall row. The result of this function is a two-tuple. It's much nicer for this to be a named tuple.

Here's a named tuple for each pair:

```
from typing import NamedTuple

class Pair(NamedTuple):
    x: float
    y: float
```

We'll introduce a transformation to transform anonymous tuples into named tuples:

```
from typing import Callable, List, Tuple, Iterable
RawPairIter = Iterable[Tuple[float, float]]

pairs: Callable[[RawPairIter], List[Pair]] \
    = lambda source: list(Pair(*row) for row in source)
```

The `RawPairIter` type definition describes the intermediate output from the `series()` function. This function emits an iterable sequence of two-tuples. The `pairs` lambda object is a callable that expects an iterable and will produce a list of `Pair` named tuples.

The following shows how the `pairs()` function and the `series()` function are used to create pairs from the original data:

```
>>> series_I = pairs(series(0, data))
>>> series_II = pairs(series(1, data))
>>> series_III = pairs(series(2, data))
>>> series_IV = pairs(series(3, data))
```

Each of these series is a `list` of `Pair` objects. Each `Pair` object has `x` and `y` attributes. The data looks as follows:

```
[Pair(x=10.0, y=8.04),
 Pair(x=8.0, y=6.95),
 ...,
 Pair(x=5.0, y=5.68)]
```

For ranking, it helps to define a composite object with the rank and the original `Pair`. A type definition for this two-tuple looks like this:

```
from typing import Tuple
RankedPair = Tuple[int, Pair]
```

The `Pair` definition is the named tuple, defined previously. The `RankedPair` is a type alias for a two-tuple consisting of an integer and a `Pair` object.

Here's a generator function that will transform an iterable collection of `Pairs` into `RankedPairs`:

```
from typing import Iterable, Iterator
def rank_y(pairs: Iterable[Pair]) -> Iterator[RankedPair]:
    return enumerate(sorted(pairs, key=lambda p: p.y))
```

This applies the `enumerate()` function to create an Iterator over `RankedPair` objects. The order is based on the y attribute of the `Pair` object. Each `Pair` is wrapped in a two-tuple with the rank and the original object.

There's a more complex variation of this idea in the following example:

```
Rank2Pair = Tuple[int, RankedPair]
def rank_x(
        ranked_pairs: Iterable[RankedPair]
    ) -> Iterator[Rank2Pair]:
    return enumerate(
        sorted(ranked_pairs, key=lambda rank: rank[1].x)
    )
```

This will wrap each `RankedPair` object to create a new `Rank2Pair` object. This secondary wrapping creates two-tuples that contain two-tuples. This complex structure shows why it can be helpful to use type aliases to provide hints on the type of data being processed.

The results of `y_rank = list(rank_y(series_I))` will look as follows:

```
[(0, Pair(x=8.0, y=5.25)),
 (1, Pair(x=8.0, y=5.56)),
 ...,
 (10, Pair(x=19.0, y=12.5))
]
```

In order to perform correlation, it's necessary to apply the `rank_x()` function, as well as the `rank_y()` function. The value of `xy_rank = list(rank_x(y_rank))` will be a list of deeply nested objects, such as the following:

```
[(0, (0, Pair(x=4.0, y=4.26))),
 (1, (2, Pair(x=5.0, y=5.68))),
 ...,
 (10, (9, Pair(x=14.0, y=9.96)))
]
```

It's now possible to compute rank-order correlations between the two variables by using the *x* and *y* rankings instead of the original values in the `Pair` object.

Two complex expressions are required to extract the two rankings. For each ranked sample in the dataset, `r`, we have to compare `r[0]` with `r[1][0]`. These are the unwrap functions that undo the wrapping done previously. These are sometimes called **selector functions** because they select an item from the middle of a complex structure.

To overcome awkward references to `r[0]` and `r[1][0]`, we can write selector functions as follows:

```
x_rank = lambda ranked: ranked[0]
y_rank = lambda ranked: ranked[1][0]
raw = lambda ranked: ranked[1][1]
```

This allows us to compute correlation using `x_rank(r)` and `y_rank(r)`, making references to values less awkward.

The overall strategy involved two operations—wrapping and unwrapping. The `rank_x()` and `rank_y()` functions wrapped `Pair` objects, creating new tuples with the ranking and the original value. We've avoided stateful class definitions to create complex data structures incrementally.

Why create deeply nested tuples? The answer is simple: laziness. The processing required to unpack a `tuple` and build a new, flat `tuple` is time consuming. There's less processing involved in `wrapping` an existing `tuple`. Using a flatter structure can make subsequent processing more clear. This leads to two improvements we'd like to make; they are as follows:

- We'd like a flatter data structure. The use of type hints for `rank_x()` and `rank_y()` show this complexity. One iterates over `Tuple[int, Pair]`, the other iterates over `Tuple[int, RankedPair]`.
- The `enumerate()` function doesn't deal with ties properly. If two observations have the same value, they should get the same rank. The general rule is to average the positions of equal observations. The sequence `[0.8, 1.2, 1.2, 2.3, 18]` should have rank values of `1, 2.5, 2.5, 4`. The two ties in positions 2 and 3 have the midpoint value of `2.5` as their common rank.

We'll look closely at these two optimizations by writing a smarter rank function.

Assigning statistical ranks

We'll break the rank ordering problem into two parts. First, we'll look at a generic, higher-order function that we can use to assign ranks to either the x or y value of a `Pair` object. Then, we'll use this to create a `wrapper` around the `Pair` object that includes both x and y rankings. This will avoid a deeply nested structure.

The following is a function that will create a rank order for each observation in a dataset:

```
from typing import Callable, Tuple, List, TypeVar, cast, Dict
D_ = TypeVar("D_")
K_ = TypeVar("K_")
def rank(
        data: Iterable[D_],
        key: Callable[[D_], K_]=lambda obj: cast(K_, obj)
    ) -> Iterator[Tuple[float, D_]]:

    def build_duplicates(
            duplicates: Dict[K_, List[D_]],
            data_iter: Iterator[D_],
            key: Callable[[D_], K_]
        ) -> Dict[K_, List[D_]]:
        for item in data_iter:
            duplicates[key(item)].append(item)
        return duplicates

    def rank_output(
            duplicates: Dict[K_, List[D_]],
            key_iter: Iterator[K_],
            base: int=0
        ) -> Iterator[Tuple[float, D_]]:
        for k in key_iter:
            dups = len(duplicates[k])
            for value in duplicates[k]:
                yield (base+1+base+dups)/2, value
            base += dups

    duplicates = build_duplicates(
        defaultdict(list), iter(data), key)
    return rank_output(duplicates, iter(sorted(duplicates)), 0)
```

This rank ordering function has two internal functions to transform a list of items to a list of two-tuples with rank and the original item. The first step is the `build_duplicates()` function, which creates a dictionary, `duplicates`, to map each key value to a sequence of items that share that value. The second step is the `rank_output()` function, which emits a sequence of two-tuples based on the duplicates dictionary.

To clarify the relationships, there are two type variables used. The `D_` type variable represents the original data type. For example, this might be a `Leg` object, or any other complex object. The `K_` type variable is the key used for rank ordering. This can be a distinct type, for example, the `float` distance value extracted from a given `Leg` named tuple. The given key function performs this transformation from data item to key item, the type hint for this is `Callable[[D_], K_]`.

The `build_duplicates()` function works with a stateful object to build the dictionary that maps keys to values. This implementation relies on Tail-Call Optimization of a recursive algorithm. The arguments to `build_duplicates()` expose the internal state as argument values. A base case for the recursion is when `data_iter` is empty, and `base` is zero. These variables aren't necessary for the iterative version, but they can be helpful for visualizing how the recursion would look.

Similarly, the `rank_output()` function could be defined recursively to emit the original collection of values as two-tuples with the assigned rank values. What's shown is an optimized version with two nested `for` loops. To make the rank value computation explicit, it includes the low end of the range (`base+1`), the high end of the range (`base+dups`), and computes the midpoint of these two values. If there is only a single `duplicate`, the rank value is `(2*base+2)/2`, which has the advantage of being a general solution.

The dictionary of duplicates has the type hint of `Dict[K_, List[D_]]`, because it maps the computed key type, `K_`, to lists of the original data item type, `List[D_]`. This appears several times and a proper type variable would be appropriate to emphasize this reuse of a common type.

The following is how we can test this to be sure it works. The first example ranks individual values. The second example ranks a list of pairs, using a lambda to pick the key value from each pair:

```
>>> list(rank([0.8, 1.2, 1.2, 2.3, 18]))
[(1.0, 0.8), (2.5, 1.2), (2.5, 1.2), (4.0, 2.3), (5.0, 18)]
>>> data= [(2, 0.8), (3, 1.2), (5, 1.2), (7, 2.3), (11, 18)]
>>> list(rank(data, key=lambda x:x[1]))
[(1.0, (2, 0.8)),
```

```
    (2.5, (3, 1.2)),
    (2.5, (5, 1.2)),
    (4.0, (7, 2.3)),
    (5.0, (11, 18))]
```

The sample data included two identical values. The resulting ranks split positions 2 and 3 to assign position `2.5` to both values. This is the common statistical practice for computing the Spearman rank-order correlation between two sets of values.

The `rank()` function involves rearranging the input data as part of discovering duplicated values. If we want to rank on both the `x` and `y` values in each pair, we need to reorder the data twice.

Wrapping instead of state changing

We have two general strategies to do wrapping; they are as follows:

- **Parallelism**: We can create two copies of the data and rank each copy. We then need to reassemble the two copies into a final result that includes both rankings. This can be a bit awkward because we'll need to somehow merge two sequences that are likely to be in different orders.
- **Serialism**: We can compute ranks on one variable and save the results as a wrapper that includes the original raw data. We can then rank this wrapped data on the other variable. While this can create a complex structure, we can optimize it slightly to create a flatter wrapper for the final results.

The following is how we can create an object that wraps a pair with the rank order based on the `y` value:

```
from typing import NamedTuple
class Ranked_Y(NamedTuple):
    r_y: float
    raw: Pair

def rank_y(pairs: Iterable[Pair]) -> Iterable[Ranked_Y]:
    return (
        Ranked_Y(rank, data)
        for rank, data in rank(pairs, lambda pair: pair.y)
    )
```

We've defined a NamedTuple subclass, Rank_Y, that contains the y ranking plus the original, raw, value. Our rank_y() function will create instances of this tuple by applying the rank() function using a lambda that selects the y value of each pairs object. We then created instances of the resulting two tuples.

The idea is that we can provide the following input:

```
>>> data = (Pair(x=10.0, y=8.04),
...     Pair(x=8.0, y=6.95),
...     Pair(x=13.0, y=7.58),
etc.
...      Pair(x=5.0, y=5.68))
```

We can get the following output:

```
>>> list(rank_y(data))
[Ranked_Y(r_y=1.0, raw=Pair(x=4.0, y=4.26)),
 Ranked_Y(r_y=2.0, raw=Pair(x=7.0, y=4.82)),
 Ranked_Y(r_y=3.0, raw=Pair(x=5.0, y=5.68)),
etc.
 Ranked_Y(r_y=11.0, raw=Pair(x=12.0, y=10.84))]
```

The raw Pair objects have been wrapped to create a new Ranked_Y object that includes the rank. This isn't all we need; we'll need to wrap this one more time to create an object that has both x and y rank information.

Rewrapping instead of state changing

We can use a NamedTuple subclass named Ranked_XY that contains two attributes: r_x and ranked_y. The ranked_y attribute is an instance of Ranked_Y that has two attributes: r_y and raw. Although this is very easy to build, the resulting objects are annoying to work with because the r_x and r_y values aren't simple peers in a flat structure. We'll introduce a slightly more complex wrapping process that produces a slightly simpler result.

We want the output to be instances of a class defined like this:

```
class Ranked_XY(NamedTuple):
    r_x: float
    r_y: float
    raw: Pair
```

We're going to create a flat `NamedTuple` with multiple peer attributes. This kind of expansion is often easier to work with than deeply nested structures. In some applications, we might have a number of transformations. For this application, we have only two transformations—**x ranking** and **y ranking**. We'll break this into two steps. First, we'll look at a simplistic wrapping, such as the one shown previously, and then a more general unwrap-rewrap.

The following is how the `x-y` ranking builds on the y-ranking:

```
def rank_xy(pairs: Sequence[Pair]) -> Iterator[Ranked_XY]:
    return (
        Ranked_XY(
            r_x=r_x, r_y=rank_y_raw[0], raw=rank_y_raw[1])
        for r_x, rank_y_raw in
            rank(rank_y(pairs), lambda r: r.raw.x)
    )
```

We've used the `rank_y()` function to build `Rank_Y` objects. Then, we applied the `rank()` function to those objects to order them by the original `x` values. The result of the second rank function will be two tuples with `(0)` the `x` rank, and `(1)` the `Rank_Y` object. We build a `Ranked_XY` object from the `x` ranking (`r_x`), the `y` ranking (`rank_y_raw[0]`), and the original object (`rank_y_raw[1]`).

What we've shown in this second function is a more general approach to adding data to a `tuple`. The construction of the `Ranked_XY` object shows how to unwrap the values from the data and rewrap to create a second, more complete structure. This approach can be used generally to introduce new variables to a `tuple`.

The following is some sample data:

```
>>> data = (Pair(x=10.0, y=8.04), Pair(x=8.0, y=6.95),
... Pair(x=13.0, y=7.58), Pair(x=9.0, y=8.81),
etc.
... Pair(x=5.0, y=5.68))
```

This allows us to create ranking objects as follows:

```
>>> list(rank_xy(data))
[Ranked_XY(r_x=1.0, r_y=1.0, raw=Pair(x=4.0, y=4.26)),
 Ranked_XY(r_x=2.0, r_y=3.0, raw=Pair(x=5.0, y=5.68)),
 Ranked_XY(r_x=3.0, r_y=5.0, raw=Pair(x=6.0, y=7.24)),
etc.
 Ranked_XY(r_x=11.0, r_y=10.0, raw=Pair(x=14.0, y=9.96))]
```

Once we have this data with the appropriate `x` and `y` rankings, we can compute the Spearman rank-order correlation value. We can compute the Pearson correlation from the raw data.

Our multiranking approach involves decomposing a `tuple` and building a new, flat `tuple` with the additional attributes we need. We will often need this kind of design when computing multiple derived values from source data.

Computing Spearman rank-order correlation

The Spearman rank-order correlation is a comparison between the rankings of two variables. It neatly bypasses the magnitude of the values, and it can often find a correlation even when the relationship is not linear. The formula is as follows:

$$\rho = 1 - \frac{6\sum(r_x - r_y)^2}{n(n^2 - 1)}$$

This formula shows us that we'll be summing the differences in rank r_x, and r_y, for all of the pairs of observed values. The Python version of this depends on the `sum()` and `len()` functions, as follows:

```
def rank_corr(pairs: Sequence[Pair]) -> float:
    ranked = rank_xy(pairs)
    sum_d_2 = sum((r.r_x - r.r_y)**2 for r in ranked)
    n = len(pairs)
    return 1-6*sum_d_2/(n*(n**2-1))
```

We've created `Rank_XY` objects for each `Pair` object. Given this, we can then subtract the `r_x` and `r_y` values from those pairs to compare their difference. We can then square and sum the differences.

A good article on statistics will provide detailed guidance on what the coefficient means. A value around `0` means that there is no correlation between the data ranks of the two series of data points. A scatter plot shows a random scattering of points. A value around `+1` or `-1` indicates a strong relationship between the two values. A graph of the pairs would show a clear line or simple curve.

The following is an example based on Anscombe's quartet series:

```
>>> data = (Pair(x=10.0, y=8.04), Pair(x=8.0, y=6.95),
... Pair(x=13.0, y=7.58), Pair(x=9.0, y=8.81),
```

```
... Pair(x=11.0, y=8.33), Pair(x=14.0, y=9.96),
... Pair(x=6.0, y=7.24), Pair(x=4.0, y=4.26),
... Pair(x=12.0, y=10.84), Pair(x=7.0, y=4.82),
... Pair(x=5.0, y=5.68))
>>> round(pearson_corr( data ), 3)
0.816
```

For this particular dataset, the correlation is strong.

In `Chapter 4`, *Working with Collections*, we showed how to compute the Pearson correlation coefficient. The function we showed, `corr()`, worked with two separate sequences of values. We can use it with our sequence of `Pair` objects as follows:

```
import Chapter_4.ch04_ex4
def pearson_corr(pairs: Sequence[Pair]) -> float:
    X = tuple(p.x for p in pairs)
    Y = tuple(p.y for p in pairs)
    return ch04_ex4.corr(X, Y)
```

We've unwrapped the `Pair` objects to get the raw values that we can use with the existing `corr()` function. This provides a different correlation coefficient. The Pearson value is based on how well the standardized values compare between two sequences. For many datasets, the difference between the Pearson and Spearman correlations is relatively small. For some datasets, however, the differences can be quite large.

To see the importance of having multiple statistical tools for exploratory data analysis, compare the Spearman and Pearson correlations for the four sets of data in the Anscombe's quartet.

Polymorphism and type-pattern matching

Some functional programming languages offer some clever approaches to the problem of working with statically typed function definitions. The problem is that many functions we'd like to write are entirely generic with respect to data type. For example, most of our statistical functions are identical for `int` or `float` numbers, as long as the division returns a value that is a subclass of `numbers.Real` (for example, `Decimal`, `Fraction`, or `float`). In many functional languages, sophisticated type or type-pattern matching rules are used by the compiler to make a single generic definition work for multiple data types. Python doesn't have this problem and doesn't need the pattern matching.

Instead of the (possibly) complex features of statically typed functional languages, Python changes the approach dramatically. Python uses dynamic selection of the final implementation of an operator based on the data types being used. In Python, we always write generic definitions. The code isn't bound to any specific data type. The Python runtime will locate the appropriate operations based on the types of the actual objects in use. The *3.3.7 Coercion rules* section of the language reference manual and the `numbers` module in the library provide details on how this mapping from operation to special method name works.

This means that the compiler doesn't certify that our functions are expecting and producing the proper data types. We generally rely on unit testing and the **mypy** tool for this kind of type checking.

In rare cases, we might need to have different behavior based on the types of data elements. We have two ways to tackle this:

- We can use the `isinstance()` function to distinguish the different cases
- We can create our own subclass of `numbers.Number` or `NamedTuple` and implement proper polymorphic special method names.

In some cases, we'll actually need to do both so that we can include appropriate data type conversions for each operation. Additionally, we'll also need to use the `cast()` function to make the types explicit to the **mypy** tool.

The ranking example in the previous section is tightly bound to the idea of applying rank-ordering to simple pairs. While this is the way the Spearman correlation is defined, a multivariate dataset have a need to do rank-order correlation among all the variables.

The first thing we'll need to do is generalize our idea of rank-order information. The following is a `NamedTuple` value that handles a `tuple` of ranks and a `raw` data object:

```
from typing import NamedTuple, Tuple, Any
class Rank_Data(NamedTuple):
    rank_seq: Tuple[float]
    raw: Any
```

A typical use of this kind of class definition is shown in this example:

```
>>> data = {'key1': 1, 'key2': 2}
>>> r = Rank_Data((2, 7), data)
>>> r.rank_seq[0]
2
>>> r.raw
{'key1': 1, 'key2': 2}
```

The row of raw data in this example is a dictionary. There are two rankings for this particular item in the overall list. An application can get the sequence of rankings as well as the original raw data item.

We'll add some syntactic sugar to our ranking function. In many previous examples, we've required either an iterable or a concrete collection. The `for` statement is graceful about working with either one. However, we don't always use the `for` statement, and for some functions, we've had to explicitly use `iter()` to make an `iterable` out of a collection. We can handle this situation with a simple `isinstance()` check, as shown in the following code snippet:

```
def some_function(seq_or_iter: Union[Sequence, Iterator]):
    if isinstance(seq_or_iter, Sequence):
        yield from some_function(iter(seq_or_iter), key)
        return
    # Do the real work of the function using the Iterator
```

This example includes a type check to handle the small difference between a `Sequence` object and an `Iterator`. Specifically, the function uses `iter()` to create an `Iterator` from a `Sequence`, and calls itself recursively with the derived value.

For rank-ordering, the `Union[Sequence, Iterator]` will be supported. Because the source data must be sorted for ranking, it's easier to use `list()` to transform a given iterator into a concrete sequence. The essential `isinstance()` check will be used, but instead of creating an iterator from a sequence (as shown previously), the following examples will create a sequence object from an iterator.

In the context of our rank-ordering function, we can make the function somewhat more generic. The following two expressions define the inputs:

```
Source = Union[Rank_Data, Any]
Union[Sequence[Source], Iterator[Source]]
```

There are four combinations defined by these two types:

- `Sequence[Rank_Data]`
- `Sequence[Any]`
- `Iterator[Rank_Data]`
- `Iterator[Any]`

Here's the `rank_data()` function with three cases for handling the four combinations of data types:

```
from typing import (
    Callable, Sequence, Iterator, Union, Iterable,
    TypeVar, cast, Union
)
K_ = TypeVar("K_") # Some comparable key type used for ranking.
Source = Union[Rank_Data, Any]
def rank_data(
        seq_or_iter: Union[Sequence[Source], Iterator[Source]],
        key: Callable[[Rank_Data], K_] = lambda obj: cast(K_, obj)
    ) -> Iterable[Rank_Data]:

    if isinstance(seq_or_iter, Iterator):
        # Iterator? Materialize a sequence object
        yield from rank_data(list(seq_or_iter), key)
        return

    data: Sequence[Rank_Data]
    if isinstance(seq_or_iter[0], Rank_Data):
        # Collection of Rank_Data is what we prefer.
        data = seq_or_iter
    else:
        # Convert to Rank_Data and process.
        empty_ranks: Tuple[float] = cast(Tuple[float], ())
        data = list(
            Rank_Data(empty_ranks, raw_data)
            for raw_data in cast(Sequence[Source], seq_or_iter)
        )

    for r, rd in rerank(data, key):
        new_ranks = cast(
            Tuple[float],
            rd.rank_seq + cast(Tuple[float], (r,)))
        yield Rank_Data(new_ranks, rd.raw)
```

We've decomposed the ranking into three cases to cover the four different types of data. The following are the cases defined by the union of unions:

- Given an `Iterator` (an object without a usable `__getitem__()` method), we'll materialize a `list` object to work with. This will work for `Rank_Data` as well as any other raw data type. This case covers objects which are `Iterator[Rank_Data]` as well as `Iterator[Any]`.

- Given a `Sequence[Any]`, we'll wrap the unknown objects into `Rank_Data` tuples with an empty collection of rankings to create a `Sequence[Rank_Data]`.
- Finally, given a `Sequence[Rank_Data]`, add yet another ranking to the tuple of ranks inside the each `Rank_Data` container.

The first case calls `rank_data()` recursively. The other two cases both rely on a `rerank()` function that builds a new `Rank_Data` tuple with additional ranking values. This contains several rankings for a complex record of raw data values.

Note that a relatively complex `cast()` expression is required to disambiguate the use of generic tuples for the rankings. The **mypy** tool offers a `reveal_type()` function that can be incorporated to debug the inferred types.

The `rerank()` function follows a slightly different design to the example of the `rank()` function shown previously. It yields two-tuples with the rank and the original data object:

```
def rerank(
        rank_data_iter: Iterable[Rank_Data],
        key: Callable[[Rank_Data], K_]
    ) -> Iterator[Tuple[float, Rank_Data]]:
    sorted_iter = iter(
        sorted(
            rank_data_iter, key=lambda obj: key(obj.raw)
        )
    )
    # Apply ranker to head, *tail = sorted(rank_data_iter)
    head = next(sorted_iter)
    yield from ranker(sorted_iter, 0, [head], key)
```

The idea behind `rerank()` is to sort a collection of `Rank_Data` objects. The first item, `head`, is used to provide a seed value to the `ranker()` function. The `ranker()` function can examine the remaining items in the iterable to see if they match this initial value, this allows computing a proper rank for a batch of matching items.

The `ranker()` function accepts a sorted iterable of data, a base rank number, and an initial collection of items of the minimum rank. The result is an iterable sequence of two-tuples with a rank number and an associated `Rank_Data` object:

```
def ranker(
        sorted_iter: Iterator[Rank_Data],
        base: float,
        same_rank_seq: List[Rank_Data],
        key: Callable[[Rank_Data], K_]
    ) -> Iterator[Tuple[float, Rank_Data]]:
```

```
        try:
            value = next(sorted_iter)
        except StopIteration:
            dups = len(same_rank_seq)
            yield from yield_sequence(
               (base+1+base+dups)/2, iter(same_rank_seq))
            return
        if key(value.raw) == key(same_rank_seq[0].raw):
            yield from ranker(
                sorted_iter, base, same_rank_seq+[value], key)
        else:
            dups = len(same_rank_seq)
            yield from yield_sequence(
                (base+1+base+dups)/2, iter(same_rank_seq))
            yield from ranker(
                sorted_iter, base+dups, [value], key)
```

This starts by attempting to extract the next item from the `sorted_iter` collection of sorted `Rank_Data` items. If this fails with a `StopIteration` exception, there is no next item, the source was exhausted. The final output is the final batch of equal-valued items in the `same_rank_seq` sequence.

If the sequence has a next item, the `key()` function extracts the key value. If this new value matches the keys in the `same_rank_seq` collection, it is accumulated into the current batch of same-valued keys. The final result is based on the rest of the items in `sorted_iter`, the current value for the rank, a larger batch of `same_rank` items that now includes the `head` value, and the original `key()` function.

If the next item's key doesn't match the current batch of equal-valued items, the final result has two parts. The first part is the batch of equal-valued items accumulated in `same_rank_seq`. This is followed by the reranking of the remainder of the sorted items. The base value for these is incremented by the number of equal-valued items, a fresh batch of equal-rank items is initialized with the distinct key, and the original `key()` extraction function is provided.

The output from `ranker()` depends on the `yield_sequence()` function, which looks as follows:

```
    def yield_sequence(
            rank: float,
            same_rank_iter: Iterator[Rank_Data]
        ) -> Iterator[Tuple[float, Rank_Data]]:
        head = next(same_rank_iter)
        yield rank, head
        yield from yield_sequence(rank, same_rank_iter)
```

We've written this in a way that emphasizes the recursive definition. For any practical work, this should be optimized into a single `for` statement.

When doing Tail-Call Optimization to transform a recursion into a loop define unit test cases first. Be sure the recursion passes the unit test cases before optimizing.

The following are some examples of using this function to rank (and rerank) data. We'll start with a simple collection of scalar values:

```
>>> scalars= [0.8, 1.2, 1.2, 2.3, 18]
>>> list(rank_data(scalars))
[Rank_Data(rank_seq=(1.0,), raw=0.8),
 Rank_Data(rank_seq=(2.5,), raw=1.2),
 Rank_Data(rank_seq=(2.5,), raw=1.2),
 Rank_Data(rank_seq=(4.0,), raw=2.3),
 Rank_Data(rank_seq=(5.0,), raw=18)]
```

Each value becomes the `raw` attribute of a `Rank_Data` object.

When we work with a slightly more complex object, we can also have multiple rankings. The following is a sequence of two tuples:

```
>>> pairs = ((2, 0.8), (3, 1.2), (5, 1.2), (7, 2.3), (11, 18))
>>> rank_x = list(rank_data(pairs, key=lambda x:x[0]))
>>> rank_x
[Rank_Data(rank_seq=(1.0,), raw=(2, 0.8)),
Rank_Data(rank_seq=(2.0,), raw=(3, 1.2)),
Rank_Data(rank_seq=(3.0,), raw=(5, 1.2)),
Rank_Data(rank_seq=(4.0,), raw=(7, 2.3)),
Rank_Data(rank_seq=(5.0,), raw=(11, 18))]
```

```
>>> rank_xy = list(rank_data(rank_x, key=lambda x:x[1] ))
>>> rank_xy
[Rank_Data(rank_seq=(1.0, 1.0), raw=(2, 0.8)),
Rank_Data(rank_seq=(2.0, 2.5), raw=(3, 1.2)),
Rank_Data(rank_seq=(3.0, 2.5), raw=(5, 1.2)),
Rank_Data(rank_seq=(4.0, 4.0), raw=(7, 2.3)),
Rank_Data(rank_seq=(5.0, 5.0), raw=(11, 18))]
```

Here, we defined a collection of pairs. Then, we ranked the two tuples, assigning the sequence of `Rank_Data` objects to the `rank_x` variable. We then ranked this collection of `Rank_Data` objects, creating a second rank value and assigning the result to the `rank_xy` variable.

The resulting sequence can be used for a slightly modified `rank_corr()` function to compute the rank correlations of any of the available values in the `rank_seq` attribute of the `Rank_Data` objects. We'll leave this modification as an exercise for the reader.

Summary

In this chapter, we looked at different ways to use `NamedTuple` objects to implement more complex data structures. The essential features of a `NamedTuple` are a good fit with functional design. They can be created with a creation function and accessed by position as well as name.

We looked at how to use immutable `NamedTuple` objects instead of stateful object definitions. The core technique for replacing state changes is to wrap objects in larger `tuple` objects.

We also looked at ways to handle multiple data types in Python. For most arithmetic operations, Python's internal method dispatch locates proper implementations. To work with collections, however, we might want to handle iterators and sequences slightly differently.

In the next two chapters, we'll look at the `itertools` module. This library module provides a number of functions that help us work with iterators in sophisticated ways. Many of these tools are examples of higher-order functions. They can help make a functional design stay succinct and expressive.

8
The Itertools Module

Functional programming emphasizes stateless objects. In Python, this leads us to work with generator expressions, generator functions, and iterables, instead of large, mutable, collection objects. In this chapter, we'll look at elements of the `itertools` library. This library has numerous functions to help us work with iterable sequences of objects, as well as collection objects.

We introduced iterator functions in `Chapter 3`, *Functions, Iterators, and Generators*. In this chapter, we'll expand on that superficial introduction. We used some related functions in `Chapter 5`, *Higher-Order Functions*.

These functions behave as if they are proper, lazy, Python iterables. Some of them create intermediate objects, however; this leads to them consuming a large amount of memory. Since implementations may change with Python releases, we can't provide function-by-function advice here. If you have performance or memory issues, ensure that you check the implementation.

"Use the source, Luke" is common advice.

There are a large number of iterator functions in the `itertools` module. We'll examine some of the functions in the next chapter. In this chapter, we'll look at three broad groupings of iterator functions. These are as follows:

- Functions that work with infinite iterators. These can be applied to any iterable or an iterator over any collection. For example, the `enumerate()` function doesn't require an upper bound on the items in the iterable.
- Functions that work with finite iterators. These can either accumulate a source multiple times, or they produce a reduction of the source. For example, grouping the items produced by an iterator requires an upper bound.

- The `tee` iterator function clones an iterator into several copies that can each be used independently. This provides a way to overcome the primary limitation of Python iterators: they can be used only once.

We need to emphasize the important limitation of iterables that we've touched upon in other places: they can only be used once.

> Iterables can be used only once.
>
> This can be astonishing because there's no error. Once exhausted, they appear to have no elements and will raise the `StopIteration` exception every time they're used.

There are some other features of iterators that aren't such profound limitations. They are as follows:

- There's no `len()` function for an iterable.
- Iterables can do `next()` operations, unlike a container. We can create an iterator for a container using the `iter()` function; this object has a `next()` operation.
- The `for` statement makes the distinction between containers and iterables invisible by using the `iter()` function automatically. A container object, for example, a `list`, responds to this function by producing an iterator over the items. An iterable object, for example, a generator function, simply returns itself, since it follows the iterator protocol.

These points will provide some necessary background for this chapter. The idea of the `itertools` module is to leverage what iterables can do to create succinct, expressive applications without the complex-looking overheads associated with the details of managing the iterables.

Working with the infinite iterators

The `itertools` module provides a number of functions that we can use to enhance or enrich an iterable source of data. We'll look at the following three functions:

- `count()`: This is an unlimited version of the `range()` function
- `cycle()`: This will reiterate a cycle of values
- `repeat()`: This can repeat a single value an indefinite number of times

Our goal is to understand how these various iterator functions can be used in generator expressions and with generator functions.

Counting with count()

The built-in `range()` function is defined by an upper limit: the lower limit and step values are optional. The `count()` function, on the other hand, has a start and optional step, but no upper limit.

This function can be thought of as the primitive basis for a function such as `enumerate()`. We can define the `enumerate()` function in terms of `zip()` and `count()` functions, as follows:

```
enumerate = lambda x, start=0: zip(count(start), x)
```

The `enumerate()` function behaves as if it's a `zip()` function that uses the `count()` function to generate the values associated with some iterator.

Consequently, the following two commands are equivalent to each other:

```
>>> list(zip(count(), iter('word')))
[(0, 'w'), (1, 'o'), (2, 'r'), (3, 'd')]
>>> list(enumerate(iter('word')))
[(0, 'w'), (1, 'o'), (2, 'r'), (3, 'd')]
```

Both will emit a sequence of numbers of two tuples. The first item in each tuple is an integer counter. The second item comes from the iterator. In this example, the iterator is built from a string of characters.

The `zip()` function is made slightly simpler with the use of the `count()` function, as shown in the following command:

```
zip(count(1,3), some_iterator)
```

The value of `count(b, s)` is the sequence of values $\{b, b+f, b+2f, b+3f, \ldots\}$. In this example, it will provide values of 1, 4, 7, 10, and so on as the identifiers for each value from the enumerator. A sequence such as this is a challenge with the `enumerate()` function because it doesn't provide a way to change the step. Here's how this can be done with the `enumerate()` function:

```
((1+3*e, x) for e,x in enumerate(some_iterator))
```

Counting with float arguments

The `count()` function permits non-integer values. We can use something such as the `count(0.5, 0.1)` method to provide floating-point values. This will accumulate an error if the increment value doesn't have an exact representation. It's generally better to use integer `count()` arguments such as `(0.5+x*.1 for x in count())` to assure that representation errors don't accumulate.

Here's a way to examine the accumulating error. This exploration of the `float` approximation shows some interesting functional programming techniques.

We'll define a function that will evaluate items from an iterator until some condition is met. Here's how we can define the `until()` function:

```
from typing import Callable, Iterator TypeVar
T_ = TypeVar("T_")
def until(
        terminate: Callable[[T_], bool],
        iterator: Iterator[T_]
    ) -> T_:
    i = next(iterator)
    if terminate(i):
        return i
    return until(terminate, iterator)
```

This function starts by getting the next value from the iterator object. The type is associated with the type variable, `T_`. If the chosen item passes the test, that is, the desired value, iteration stops and the return value will be of the given type associated with the type variable, `T_`. Otherwise, we'll evaluate this function recursively to search for a subsequent value that passes the test.

Here is an example of an iterable object and a comparison function:

```
Generator = Iterator[Tuple[float, float]]
source: Generator = zip(count(0, 0.1), (.1*c for c in count()))

Extractor = Callable[[Tuple[float, float]], float]
x: Extractor = lambda x_y: x_y[0]
y: Extractor = lambda x_y: x_y[1]

Comparator = Callable[[Tuple[float, float]], bool]
neq: Comparator = lambda xy: abs(x(xy)-y(xy)) > 1.0E-12
```

The generator, `source`, has provided a type hint on the assignment statement to show that it iterates over two-tuples. The two extractor functions, `x()`, and `y()`, decompose a two-tuple to create a float result. The comparator function, `neq()`, returns a Boolean result given a tuple of float values.

The lambda objects are created with assignment statements. The type hints were used to clarify the parameter and result types for the **mypy** tool.

When checking the `until()` function, the **mypy** tool will associate the type variable, `T_`, with the concrete type of `Tuple[float, float]`. This association will confirm that the `source` generator and `neq()` function will work with the `until()` function.

When we evaluate the `until(neq, source)` method, we'll repeatedly compare float approximations of decimal values until they differ. One approximation, computed by `count()` is a sum of 1: $\sum_{x \in \mathbb{N}} .1$. The other approximation, computed by a generator, is $.1 \times \sum_{x \in \mathbb{N}} 1$. In the abstract, there's no distinction. With concrete approximations of the abstract numbers, the two values will differ.

The result is as follows:

```
>>> until(neq, source)
(92.799999999999, 92.80000000000001)
```

After 928 iterations, the sum of the error bits has accumulated to 10^{-12}. Neither value has an exact binary representation.

The `until()` function example is close to the Python recursion limit. We'd need to rewrite the function to use tail-call optimization to locate counts with larger accumulated errors.

The smallest detectable difference can be computed as follows:

```
>>> until(lambda x, y: x != y, source)
(0.6, 0.6000000000000001)
```

This uses a simple equality check instead of an error range. After six steps, the `count(0, 0.1)` method has accumulated a measurable error of 10^{-16}. When looking at how $\frac{1}{10}$ is represented as a binary value, an infinite binary expansion would be required. This fraction is truncated, leading to a small error that can be accumulated.

Re-iterating a cycle with cycle()

The `cycle()` function repeats a sequence of values. This can be used when partitioning data into subsets by cycling among the dataset identifiers.

We can imagine using it to solve silly fizz-buzz problems. Visit `http://rosettacode.org/wiki/FizzBuzz` for a comprehensive set of solutions to a fairly trivial programming problem. Also see `https://projecteuler.net/problem=1` for an interesting variation on this theme.

We can use the `cycle()` function to emit sequences of `True` and `False` values as follows:

```
m3 = (i == 0 for i in cycle(range(3)))
m5 = (i == 0 for i in cycle(range(5)))
```

These are infinite sequences that have a pattern of `[True, False, False, True, False, False, ...]` or `[True, False, False, False, False, True, False, False, False, False, ...]`.

If we zip together a finite collection of numbers and these two derived flags, we'll get a set of three-tuples with a number, a multiple of three conditions, and a multiple of five conditions. It's important to introduce a finite iterable to create a proper upper bound on the volume of data being generated. Here's a sequence of values and their multiplier flags:

```
multipliers = zip(range(10), m3, m5)
```

This is a generator; we can use `list(multipliers)` to see the resulting object. It looks like this:

```
[(0, True, True), (1, False, False), (2, False, False), ..., (9, True,
False)]
```

We can now decompose the triples and use a filter to pass numbers that are multiples and reject all others:

```
total = sum(i
    for i, *multipliers in multipliers
    if any(multipliers)
)
```

The `for` clause decomposes each triple into two parts: the value, `i`, and the flags, `multipliers`. If any of the multipliers are true, the value is passed, otherwise, it's rejected.

This function has another, more valuable, use for exploratory data analysis.

We often need to work with samples of large sets of data. The initial phases of cleansing and model creation are best developed with small sets of data and tested with larger and larger sets of data. We can use the `cycle()` function to fairly select rows from within a larger set. Given a population size, N_p, and the desired sample size, N_s, this is the required size of the cycle that will produce appropriate subsets:

$$c = \frac{N_p}{N_s}$$

We'll assume that the data can be parsed with the `csv` module. This leads to an elegant way to create subsets. Given a value for the `cycle_size`, and two open files, `source_file` and `target_file`, we can create subsets using the following commands:

```
chooser = (x == 0 for x in cycle(range(cycle_size)))
rdr = csv.reader(source_file)
wtr = csv.writer(target_file)
wtr.writerows(
    row for pick, row in zip(chooser, rdr) if pick
)
```

We created a `cycle()` function based on the selection factor, `cycle_size`. For example, we might have a population of 10 million records: a 1,000-record subset involves a cycle size of 10,000 records. We assumed that this snippet of code is nestled securely inside a `with` statement that opens the relevant files.

We can use a simple generator expression to filter the data using the `cycle()` function and the source data that's available from the CSV reader. Since the `chooser` expression and the expression used to write the rows are both non-strict, there's little memory overhead from this kind of processing.

We can also rewrite this method to use `compress()`, `filter()`, and `islice()` functions, as we'll see later in this chapter.

This design will also reformat a file from any non-standard CSV-like format into a standardized CSV format. As long as we define parser functions that return consistently defined tuples and write consumer functions that write tuples to the target files, we can do a great deal of cleansing and filtering with relatively short, clear scripts.

Repeating a single value with repeat()

The `repeat()` function seems like an odd feature: it returns a single value over and over again. It can serve as an alternative for the `cycle()` function when a single value is needed.

The difference between selecting all of the data and selecting a subset of the data can be expressed with this. The function `(x==0 for x in cycle(range(size)))` emits a `[True, False, False, ...]` pattern, suitable for picking a subset. The function `(x==0 for x in repeat(0))` emits a `[True, True, True, ...]` pattern, suitable for selecting all of the data.

We can think of the following kinds of commands:

```
all = repeat(0)
subset = cycle(range(100))
choose = lambda rule: (x == 0 for x in rule)
# choose(all) or choose(subset) can be used
```

This allows us to make a simple parameter change, which will either pick all data or pick a subset of data. This pattern can be extended to randomize the subset chosen. The following technique adds an additional kind of choice:

```
def randseq(limit):
    while True:
        yield random.randrange(limit)
randomized = randseq(100)
```

The `randseq()` function generates a potentially infinite sequence of random numbers over a given range. This fits the pattern of `cycle()` and `repeat()`.

This allows code such as the following:

```
[v for v, pick in zip(data, choose(all)) if pick]
[v for v, pick in zip(data, choose(subset)) if pick]
[v for v, pick in zip(data, choose(randomized)) if pick]
```

Using `chose(all)`, `chose(subset)`, or `chose(randomized)` is a succinct expression of which data is selected for further analysis.

Using the finite iterators

The `itertools` module provides a number of functions that we can use to produce finite sequences of values. We'll look at 10 functions in this module, plus some related built-in functions:

- `enumerate()`: This function is actually part of the `__builtins__` package, but it works with an iterator and is very similar to other functions in the `itertools` module.
- `accumulate()`: This function returns a sequence of reductions of the input iterable. It's a higher-order function and can do a variety of clever calculations.
- `chain()`: This function combines multiple iterables serially.
- `groupby()`: This function uses a function to decompose a single iterable into a sequence of iterables over subsets of the input data.
- `zip_longest()`: This function combines elements from multiple iterables. The built-in `zip()` function truncates the sequence at the length of the shortest iterable. The `zip_longest()` function pads the shorter iterables with the given fill value.
- `compress()`: This function filters one iterable based on a second iterable of `Boolean` values.
- `islice()`: This function is the equivalent of a slice of a sequence when applied to an iterable.
- `dropwhile()` and `takewhile()`: Both of these functions use a `Boolean` function to filter items from an iterable. Unlike `filter()` or `filterfalse()`, these functions rely on a single `True` or `False` value to change their filter behavior for all subsequent values.
- `filterfalse()`: This function applies a filter function to an iterable. This complements the built-in `filter()` function.
- `starmap()`: This function maps a function to an iterable sequence of tuples using each iterable as an `*args` argument to the given function. The `map()` function does a similar thing using multiple parallel iterables.

We've grouped these functions into approximate categories. The categories are roughly related to concepts of restructuring an iterable, filtering, and mapping.

Assigning numbers with enumerate()

In `Chapter 7`, *Additional Tuple Techniques*, we used the `enumerate()` function to make a naive assignment of rank numbers to sorted data. We can do things such as pairing up a value with its position in the original sequence, as follows:

```
pairs = tuple(enumerate(sorted(raw_values)))
```

This will sort the items in `raw_values` in order, create two tuples with an ascending sequence of numbers, and materialize an object we can use for further calculations. The command and the result are as follows:

```
>>> raw_values = [1.2, .8, 1.2, 2.3, 11, 18]
>>> tuple(enumerate( sorted(raw_values)))
((0, 0.8), (1, 1.2), (2, 1.2), (3, 2.3), (4, 11), (5, 18))
```

In `Chapter 7`, *Additional Tuple Techniques*, we implemented an alternative form of enumerate, the `rank()` function, which would handle ties in a more statistically useful way.

This is a common feature that is added to a parser to record the source data row numbers. In many cases, we'll create some kind of `row_iter()` function to extract the string values from a source file. This may iterate over the `string` values in tags of an XML file or in columns of a CSV file. In some cases, we may even be parsing data presented in an HTML file parsed with Beautiful Soup.

In `Chapter 4`, *Working with Collections*, we parsed an XML file to create a simple sequence of position tuples. We then created legs with a start, end, and distance. We did not, however, assign an explicit leg number. If we ever sorted the trip collection, we'd be unable to determine the original ordering of the legs.

In `Chapter 7`, *Additional Tuple Techniques*, we expanded on the basic parser to create named tuples for each leg of the trip. The output from this enhanced parser looks as follows:

```
(Leg(start=Point(latitude=37.54901619777347, longitude=
 -76.33029518659048), end=Point(latitude=37.840832, longitude=
 -76.273834), distance=17.7246),
 Leg(start=Point(latitude=37.840832, longitude=-76.273834),
 end=Point(latitude=38.331501, longitude=-76.459503),
 distance=30.7382),
 Leg(start=Point(latitude=38.331501, longitude=-76.459503),
 end=Point(latitude=38.845501, longitude=-76.537331),
 distance=31.0756),
...,
 Leg(start=Point(latitude=38.330166, longitude=-76.458504),
 end=Point(latitude=38.976334, longitude=-76.473503),
```

```
    distance=38.8019))
```

The first `Leg` function is a short trip between two points on the Chesapeake Bay.

We can add a function that will build a more complex tuple with the input order information as part of the tuple. First, we'll define a slightly more complex version of the `Leg` class:

```
from typing import NamedTuple
class Point(NamedTuple):
    latitude: float
    longitude: float

class Leg(NamedTuple):
    start: Point
    end: Point
    distance: float
```

The `Leg` definition is similar to the `Leg` definition shown in Chapter 7, *Additional Tuple Techniques,* but it includes the order as well as the other attributes. We'll define a function that decomposes pairs and creates `Leg` instances as follows:

```
from typing import Iterator

def ordered_leg_iter(
        pair_iter: Iterator[Tuple[Point, Point]]
    ) -> Iterator[Leg]:
    for order, pair in enumerate(pair_iter):
        start, end = pair
        yield Leg(
            order,
            start,
            end,
            round(haversine(start, end), 4)
        )
```

We can use this function to enumerate each pair of starting and ending points. We'll decompose the pair and then re-assemble the `order`, `start`, and `end` parameters and the `haversine(start,end)` parameter's value as a single `Leg` instance. This `generator` function will work with an iterable sequence of pairs.

In the context of the preceding explanation, it is used as follows:

```
filename = "file:./Winter%202012-2013.kml"
with urllib.request.urlopen(filename) as source:
    path_iter = float_lat_lon(row_iter_kml(source))
```

```
pair_iter = legs(path_iter)
trip_iter = ordered_leg_iter( pair_iter )
trip = list(trip_iter)
```

We've parsed the original file into the path points, created start—end pairs, and then created a trip that was built of individual `Leg` objects. The `enumerate()` function assures that each item in the iterable sequence is given a unique number that increments from the default starting value of 0. A second argument value to the `enumerate()` function can be given to provide an alternate starting value.

Running totals with accumulate()

The `accumulate()` function folds a given function into an iterable, accumulating a series of reductions. This will iterate over the running totals from another iterator; the default function is `operator.add()`. We can provide alternative functions to change the essential behavior from sum to product. The Python library documentation shows a particularly clever use of the `max()` function to create a sequence of maximum values so far.

One application of running totals is quartiling data. We can compute the running total for each sample and divide them into quarters with an `int(4*value/total)` calculation.

In the *Assigning numbers with* `enumerate()` section, we introduced a sequence of latitude-longitude coordinates that describe a sequence of legs on a voyage. We can use the distances as a basis for quartiling the waypoints. This allows us to determine the midpoint in the trip.

The value of the `trip` variable looks as follows:

```
(Leg(start=Point(latitude=37.54901619777347, longitude=
 -76.33029518659048), end=Point(latitude=37.840832, longitude=
 -76.273834), distance=17.7246),
 Leg(start=Point(latitude=37.840832, longitude=-76.273834),
 end=Point(latitude=38.331501, longitude=-76.459503),
 distance=30.7382),
...,
 Leg(start=Point(latitude=38.330166, longitude=-76.458504),
 end=Point(latitude=38.976334, longitude=-76.473503),
 distance=38.8019))
```

Each `Leg` object has a start point, an end point, and a distance. The calculation of quartiles looks like the following code:

```
distances = (leg.distance for leg in trip)
distance_accum = tuple(accumulate(distances))
total = distance_accum[-1]+1.0
quartiles = tuple(int(4*d/total) for d in distance_accum)
```

We extracted the distance values and computed the accumulated distances for each leg. The last of the accumulated distances is the total. We've added `1.0` to the total to assure that `4*d/total` is 3.9983, which truncates to 3. Without the `+1.0`, the final item would have a value of `4`, which is an impossible fifth quartile. For some kinds of data (with extremely large values) we may have to add a larger value.

The value of the `quartiles` variable is as follows:

```
[0, 0, 0, 0, 0, 0, 0, 0, 0, 0, 0, 0, 0, 0, 0, 0, 0, 0, 0, 0, 0, 0, 0,
 1, 1, 1, 1, 1, 1, 1, 1, 1, 1, 1, 1, 1, 1,
 2, 2, 2, 2, 2, 2, 2, 2, 2, 2, 2, 2, 2, 2, 2, 2, 2, 2, 2,
 3, 3, 3, 3, 3, 3, 3, 3, 3, 3, 3, 3, 3, 3, 3, 3, 3]
```

We can use the `zip()` function to merge this sequence of quartile numbers with the original data points. We can also use functions such as `groupby()` to create distinct collections of the legs in each quartile.

Combining iterators with chain()

We can use the `chain()` function to combine a collection of iterators into a single, overall iterator. This can be helpful to combine data that was decomposed via the `groupby()` function. We can use this to process a number of collections as if they were a single collection.

In particular, we can combine the `chain()` function with the `contextlib.ExitStack()` method to process a collection of files as a single iterable sequence of values. We can do something such as this:

```
from contextlib import ExitStack
import csv
def row_iter_csv_tab(*filenames: str) -> Iterator[List[str]]:
    with ExitStack() as stack:
        files = [
            stack.enter_context(cast(TextIO, open(name, 'r')))
            for name in filenames
        ]  # type: List[TextIO]
```

```
        readers = map(
            lambda f: csv.reader(f, delimiter='\t'),
            files)
        yield from chain(*readers)
```

We've created an `ExitStack` object that can contain a number of individual contexts open. When the `with` statement finishes, all items in the `ExitStack` object will be closed properly. A sequence of open file objects is assigned to the `files` variable. These objects were also entered into the `ExitStack` object.

Given the sequence of files in the `files` variable, we created a sequence of CSV readers in the `readers` variable. In this case, all of our files have a common tab-delimited format, which makes it very pleasant to open all of the files with a simple, consistent application of a function to the sequence of files.

We could also open the files, using the following command:

```
readers = [csv.reader(f, delimiter='\t') for f in files]
```

Finally, we chained all of the readers into a single iterator with `chain(*readers)`. This was used to yield the sequence of rows from all of the files.

It's important to note that we can't return the `chain(*readers)` object. If we do, this would exit the `with` statement context, closing all the source files. Instead, we must yield individual rows from the generator so that the `with` statement context is kept active.

Partitioning an iterator with groupby()

We can use the `groupby()` function to partition an iterator into smaller iterators. This works by evaluating the given `key()` function for each item in the given iterable. If the key value matches the previous item's key, the two items are part of the same partition. If the key does not match the previous item's key, the previous partition is ended and a new partition is started.

The output from the `groupby()` function is a sequence of two tuples. Each tuple has the group's key value and an iterable over the items in the group. Each group's iterator can be preserved as a tuple or processed to reduce it to some summary value. Because of the way the group iterators are created, they can't be preserved.

In the *Running totals with* `accumulate()` section, earlier in the chapter, we showed how to compute quartile values for an input sequence.

Given the trip variable with the raw data and the quartile variable with the quartile assignments, we can group the data using the following commands:

```
group_iter = groupby(
    zip(quartile, trip),
    key=lambda q_raw: q_raw[0])
for group_key, group_iter in group_iter:
    print(group_key, tuple(group_iter))
```

This will start by zipping the quartile numbers with the raw trip data, iterating over two tuples. The groupby() function will use the given lambda variable to group by the quartile number. We used a for loop to examine the results of the groupby() function. This shows how we get a group key value and an iterator over members of the group.

The input to the groupby() function must be sorted by the key values. This will assure that all of the items in a group will be adjacent.

Note that we can also create groups using the defaultdict(list) method, as follows:

```
from collections import defaultdict
from typing import Iterable, Callable, Tuple, List, Dict

D_ = TypeVar("D_")
K_ = TypeVar("K_")
def groupby_2(
        iterable: Iterable[D_],
        key: Callable[[D_], K_]
    ) -> Iterator[Tuple[K_, Iterator[D_]]]:
    groups: Dict[K_, List[D_]] = defaultdict(list)
    for item in iterable:
        groups[key(item)].append(item)
    for g in groups:
        yield g, iter(groups[g])
```

We created a defaultdict class with a list object as the value associated with each key. The type hints clarify the relationship between the key() function, which emits objects of some arbitrary type associated with the type variable K_ and the dictionary, which uses the same type, K_, for the keys.

Each item will have the given key() function applied to create a key value. The item is appended to the list in the defaultdict class with the given key.

Once all of the items are partitioned, we can then return each partition as an iterator over the items that share a common key. This is similar to the groupby() function because the input iterator to this function isn't necessarily sorted in precisely the same order; it's possible that the groups may have the same members, but the order may differ.

The type hints clarify that the source is some arbitrary type, associated with the variable D_. The result will be an iterator that includes iterators of the type D_. This makes a strong statement that no transformation is happening: the range type matches the input domain type.

Merging iterables with zip_longest() and zip()

We saw the zip() function in Chapter 4, *Working with Collections*. The zip_longest() function differs from the zip() function in an important way: whereas the zip() function stops at the end of the shortest iterable, the zip_longest() function pads short iterables and stops at the end of the longest iterable.

The fillvalue keyword parameter allows filling with a value other than the default value, None.

For most exploratory data analysis applications, padding with a default value is statistically difficult to justify. The **Python Standard Library** document shows a few clever things that can be done with the zip_longest() function. It's difficult to expand on these without drifting far from our focus on data analysis.

Filtering with compress()

The built-in filter() function uses a predicate to determine whether an item is passed or rejected. Instead of a function that calculates a value, we can use a second, parallel iterable to determine which items to pass and which to reject.

We can think of the filter() function as having the following definition:

```
def filter(function, iterable):
    i1, i2 = tee(iterable, 2)
    return compress(i1, map(function, i2))
```

We cloned the iterable using the `tee()` function. We'll look at this function in detail later. The `map()` function will generate results of applying the filter predicate function, `function()`, to each value in the iterable, yielding a sequence of `True` and `False` values. The sequence of Booleans are used to compress the original sequence, passing only items associated with `True`. This builds the features of the `filter()` function from the more primitive features of the `compress()` function.

In the *Re-iterating a cycle with* `cycle()` section of this chapter, we looked at data selection using a simple generator expression. Its essence was as follows:

```
choose = lambda rule: (x == 0 for x in rule)
keep = [v for v, pick in zip(data, choose(all)) if pick]
```

Each value for a rule must be a function to produce a sequence of Boolean values. To choose all items, it simply repeats `True`. To pick a fixed subset, it cycles among `True` followed by $c-1$ copies of `False`.

The list comprehension can be revised as `compress(some_source, choose(rule))`. If we make that change, the processing is simplified:

```
compress(data, choose(all))
compress(data, choose(subset))
compress(data, choose(randomized))
```

These examples rely on the alternative selection rules: `all`, `subset`, and `randomized` as shown previously. The `subset` and `randomized` versions must be defined with a proper parameter to pick $\frac{1}{c}$ rows from the source. The `choose` expression will build an iterable over `True` and `False` values based on one of the selection rules. The rows to be kept are selected by applying the source iterable to the row-selection iterable.

Since all of this is non-strict, rows are not read from the source until required. This allows us to process very large sets of data efficiently. Also, the relative simplicity of the Python code means that we don't really need a complex configuration file and an associated parser to make choices among the selection rules. We have the option to use this bit of Python code as the configuration for a larger data-sampling application.

Picking subsets with islice()

In `Chapter 4`, *Working with Collections*, we looked at slice notation to select subsets from a collection. Our example was to pair up items sliced from a `list` object. The following is a simple list:

```
flat= ['2', '3', '5', '7', '11', '13', '17', '19',
  '23', '29', '31', '37', '41', '43', '47', '53',
  '59', '61', '67', '71',
...
 ]
```

We can create pairs using list slices as follows:

```
>>> list(zip(flat[0::2], flat[1::2]))
[(2, 3), (5, 7), (11, 13), ...]
```

The `islice()` function gives us similar capabilities without the overhead of materializing a `list` object. This will work with an iterable of any size. It looks like the following:

```
flat_iter_1= iter(flat)
flat_iter_2= iter(flat)
zip(
    islice(flat_iter_1, 0, None, 2),
    islice(flat_iter_2, 1, None, 2)
)
```

We created two independent iterators over a collection of data points in the `flat` variable. These could be two separate iterators over an open file or a database result set. The two iterators need to be independent so that change in one `islice()` function doesn't interfere with the other `islice()` function.

The two sets of arguments to the `islice()` function are similar to the slice notation, like `flat[0::2]` or `flat[1::2]` expressions. There's no slice-like shorthand, so the start and stop argument values are required. The step can be omitted and the default value is 1. This will produce a sequence of two tuples from the original sequence:

```
[(2, 3), (5, 7), (11, 13), (17, 19), (23, 29),
...
(7883, 7901), (7907, 7919)]
```

Since `islice()` works with an iterable, this kind of design will work with extremely large sets of data. We can use this to pick a subset out of a larger set of data. In addition to using the `filter()` or `compress()` functions, we can also use the `islice(source, 0, None, c)` method to pick a $\frac{1}{c}$ item subset from a larger set of data.

Stateful filtering with dropwhile() and takewhile()

The `dropwhile()` and `takewhile()` functions are stateful filter functions. They start in one mode; the given `predicate` function is a kind of flip-flop that switches the mode. The `dropwhile()` function starts in reject mode; when the function becomes `False`, it switches to pass mode. The `takewhile()` function starts in pass mode; when the given function becomes `False`, it switches to reject mode. Since these are filters, they will consume the entire iterable.

We can use these to skip header or footer lines in an input file. We use the `dropwhile()` function to reject header rows and pass the remaining data. We use the `takewhile()` function to pass data and reject trailer rows. We'll return to the simple GPL file format shown in `Chapter 3`, *Functions, Iterators, and Generators*. The file has a header that looks as follows:

```
GIMP Palette
Name: Crayola
Columns: 16
#
```

This is followed by rows that look like the following code:

```
255  73 108  Radical Red
```

We can easily locate the final line of the headers—the `#` line-using a parser based on the `dropwhile()` function, as follows:

```
with open("crayola.gpl") as source:
    rdr = csv.reader(source, delimiter='\t')
    rows = dropwhile(lambda row: row[0] != '#', rdr)
```

We created a CSV reader to parse the lines based on tab characters. This will neatly separate the `color` three tuple from the name. The three tuple will need further parsing. This will produce an iterator that starts with the `#` line and continues with the rest of the file.

We can use the `islice()` function to discard the first item of an iterable. We can then parse the color details as follows:

```
color_rows = islice(rows, 1, None)
colors = (
    (color.split(), name) for color, name in color_rows
)
print(list(colors))
```

The `islice(rows, 1, None)` expression is similar to asking for a `rows[1:]` slice: the first item is quietly discarded. Once the last of the heading rows have been discarded, we can parse the color tuples and return more useful color objects.

For this particular file, we can also use the number of columns located by the CSV reader function. Header rows only have a single column, allowing the use of the `dropwhile(lambda row: len(row) == 1, rdr)` expression to discard header rows. This isn't a good approach in general because locating the last line of the headers is often easier than trying to define some general pattern that distinguishes all header (or trailer) lines from the meaningful file content. In this case, the header rows were distinguishable by the number of columns; this should be considered a rarity.

Two approaches to filtering with filterfalse() and filter()

In `Chapter 5`, *Higher-Order Functions,* we looked at the built-in `filter()` function. The `filterfalse()` function from the `itertools` module could be defined from the `filter()` function, as follows:

```
filterfalse = (lambda pred, iterable:
    filter(lambda x: not pred(x), iterable)
)
```

As with the `filter()` function, the predicate function can be the `None` value. The value of the `filter(None, iterable)` method is all the `True` values in the iterable. The value of the `filterfalse(None, iterable)` method is all of the `False` values from the iterable:

```
>>> filter(None, [0, False, 1, 2])
>>> list(_)
[1, 2]

>>> filterfalse(None, [0, False, 1, 2])
<itertools.filterfalse object at 0x101b43a50>
>>> list(_)
[0, False]
```

The point of having the `filterfalse()` function is to promote reuse. If we have a succinct function that makes a filter decision, we should be able to use that function to partition input to pass as well as reject groups without having to fiddle around with logical negation.

The idea is to execute the following commands:

```
iter_1, iter_2 = iter(some_source), iter(some_source)
good = filter(test, iter_1)
bad = filterfalse(test, iter_2)
```

This will include all items from the source. The `test()` function is unchanged, and we can't introduce a subtle logic bug through improper negation of this function.

Applying a function to data via starmap() and map()

The built-in `map()` function is a higher-order function that applies a function to items from an iterable. We can think of the simple version of the `map()` function, as follows:

```
map = (lambda function, arg_iter:
    (function(a) for a in arg_iter)
)
```

This works well when the `arg_iter` parameter is an iterable that provides individual values. The actual `map()` function is a bit more sophisticated than this, and will work with a number of iterables.

The `starmap()` function in the `itertools` module is simply the `*a` version of the `map()` function. We can imagine the definition as follows:

```
starmap = (lambda function, arg_iter:
    (function(*a) for a in arg_iter)
)
```

This reflects a small shift in the semantics of the `map()` function to properly handle a tuple-of-tuples structure.

The `map()` function can also accept multiple iterables; the values from these additional iterables are zipped and it behaves like the `starmap()` function. Each zipped item from the source iterables becomes multiple arguments to the given function.

We can think of the `map(function, iter1, iter2, ..., itern)` function as having a definition like either one of the following:

```
map1 = (lambda function, *iters:
    (function(*args) for args in zip(*iters))
)
```

```
map2 = (lambda function, *iters:
    (starmap(function, zip(*iters)))
)
```

Items from each iterator are used to construct a tuple of arguments with `zip(*iters)`. This tuple is then expanded to match all of the positional parameters of the given function via the `*args` construct. We can build the `map()` function from the more general `starmap()` function.

When we look at the trip data, from the preceding commands, we can redefine the construction of a `Leg` object based on the `starmap()` function. Prior to creating `Leg` objects, we created pairs of points. Each pair looks as follows:

```
((Point(latitude=37.54901619777347, longitude=-76.33029518659048),
  Point(latitude=37.840832, longitude=-76.273834)),
...,
 (Point(latitude=38.330166, longitude=-76.458504),
  Point(latitude=38.976334, longitude=-76.473503))
)
```

We could use the `starmap()` function to assemble the `Leg` objects, as follows:

```
from Chapter_7.ch07_ex1 import float_lat_lon, Leg, Point
from Chapter_6.ch06_ex3 import row_iter_kml
from Chapter_4.ch04_ex1 import legs, haversine
from typing import List, Callable

make_leg = (lambda start, end:
    Leg(start, end,  haversine(start,end))
)  # type: Callable[[Point, Point], Leg]
with urllib.request.urlopen(url) as source:
    path_iter = float_lat_lon(row_iter_kml(source))
    pair_iter = legs(path_iter)
    trip = list(starmap(make_leg, pair_iter))
```

The `make_leg()` function accepts a pair of `Points` objects, and returns a `Leg` object with the start point, end point, and distance between the two points. The `legs()` function from `Chapter 4`, *Working With Collections,* creates pairs of `Point` objects that reflect the start and end of a leg of the voyage. The pairs created by `legs()` are provided as input to `make_leg()` to create proper `Leg` objects.

The benefit of the `starmap(function, some_list)` method is to replace a potentially wordy `(function(*args) for args in some_list)` generator expression with something a little bit simpler to read.

Cloning iterators with tee()

The `tee()` function gives us a way to circumvent one of the important Python rules for working with iterables. The rule is so important, we'll repeat it here:

Iterators can be used only once.

The `tee()` function allows us to clone an iterator. This seems to free us from having to materialize a sequence so that we can make multiple passes over the data. For example, a simple average for an immense dataset could be written in the following way:

```
def mean(iterator: Iterator[float]) -> float:
    it0, it1 = tee(iterator,2)
    N = sum(1 for x in it0)
    s1 = sum(x for x in it1)
    return s1/N
```

This would compute an average without appearing to materialize the entire dataset in memory in any form. Note that the type hint of `float` doesn't preclude integers. The **mypy** program is aware of the type coercion rules, and this definition provides a flexible way to specify that either `int` or `float` will work.

While interesting in principle, the `tee()` function's implementation suffers from a severe limitation. In most Python implementations, the cloning is done by materializing a sequence. While this circumvents the *one time only* rule for small collections, it doesn't work out well for immense collections.

Also, the current implementation of the `tee()` function consumes the source iterator. It might be nice to create some syntactic sugar to allow unlimited use of an iterator. This is difficult to manage in practice. Instead, Python obliges us to optimize the `tee()` function carefully.

The itertools recipes

Within the *itertools* chapter of the Python library documentation, there's a subsection called *Itertools Recipes*, which contains outstanding examples of ways to use the various itertools functions. Since there's no reason to reproduce these, we'll reference them here. They should be considered as required reading on functional programming in Python.

The *10.1.2* section, *Itertools Recipes* of *Python Standard Library,* is a wonderful resource. For more information visit: `https://docs.python.org/3/library/itertools.html#itertools-recipes`.

It's important to note that these aren't importable functions in the `itertools` modules. A recipe needs to be read and understood and then, perhaps, copied or modified before it's included in an application.

The following table summarizes some of the recipes that show functional programming algorithms built from the itertools basics:

Function Name	**Arguments**	**Results**
`take`	(*n*, iterable)	This returns the first *n* items of the iterable as a list. This wraps a use of `islice()` in a simple name.
`tabulate`	(function, start=0)	This returns `function(0)`, `function(1)`, and so on. This is based on a `map(function, count())`.
`consume`	(iterator, *n*)	This advances the iterator *n* steps ahead. If *n* is `None`, it consumes all of the values from the iterator.
`nth`	(iterable, n, default=None)	This returns the *n*th item or a default value. This wraps the use of `islice()` in a simple name.
`quantify`	(iterable, pred=bool)	This counts how many times the predicate is true. This uses `sum()` and `map()`, and relies on the way a Boolean predicate is effectively 1 when converted to an integer value.
`padnone`	(iterable)	This returns the sequence elements and then returns `None` indefinitely. This can create functions that behave like `zip_longest()` or `map()`.
`ncycles`	(iterable, *n*)	This returns the sequence elements *n* times.

`dotproduct`	(vec1, vec2)	This is the essential definition of a dot product. It multiplies two vectors and finds the sum of the result.
`flatten`	(listOfLists)	This flattens one level of nesting. This chains the various lists together into a single list.
`repeatfunc`	(func, times=None, *args)	This calls to `func` repeatedly with specified arguments.
`pairwise`	(iterable):	`s -> (s0,s1), (s1,s2), (s2, s3).`
`grouper`	(iterable, *n*, fillvalue=None)	This collects data into fixed-length chunks or blocks.
`roundrobin`	(*iterables)	`roundrobin('ABC', 'D', 'EF') --> A D E B F C`
`partition`	(pred, iterable)	This uses a predicate to partition entries into `False` entries and `True` entries.
`unique_everseen`	(iterable, key=None)	This lists unique elements, preserving order. It also remembers all elements ever seen.
`unique_justseen`	(iterable, key=None)	This lists unique elements, preserving order. it remembers only the element most recently seen. This is useful for deduplicating a sorted sequence.
`iter_except`	(func, exception, first=None)	This calls a function repeatedly until an exception is raised. This can be used to iterate until `KeyError` or `IndexError`.

Summary

In this chapter, we've looked at a number of functions in the `itertools` module. This library module provides a number of functions that help us to work with iterators in sophisticated ways.

We've looked at the infinite iterators; these repeat without terminating. These include the `count()`, `cycle()`, and `repeat()` functions. Since they don't terminate, the consuming function must determine when to stop accepting values.

We've also looked at a number of finite iterators. Some of these are built-in, and some of these are a part of the `itertools` module. These work with a source iterable, so they terminate when that iterable is exhausted. These functions include `enumerate()`, `accumulate()`, `chain()`, `groupby()`, `zip_longest()`, `zip()`, `compress()`, `islice()`, `dropwhile()`, `takewhile()`, `filterfalse()`, `filter()`, `starmap()`, and `map()`. These functions allow us to replace possibly complex generator expressions with simpler-looking functions.

Additionally, we looked at the recipes from the documentation, which provide yet more functions we can study and copy for our own applications. The recipes list shows a wealth of common design patterns.

In `Chapter 9`, *More Itertools Techniques*, we'll continue our study of the `itertools` module. We'll look at the iterators the focus on permutations and combinations. These don't apply when it comes to processing large sets of data. They're a different kind of iterator-based tool.

9
More Itertools Techniques

Functional programming emphasizes stateless programming. In Python, this leads us to work with generator expressions, generator functions, and iterables. In this chapter, we'll continue our study of the itertools library, with numerous functions to help us work with iterable collections.

In the previous chapter, we looked at three broad groupings of iterator functions. They are as follows:

- Functions that work with infinite iterators, which can be applied to any iterable or an iterator over any collection; they will consume the entire source
- Functions that work with finite iterators, which can either accumulate a source multiple times, or produce a reduction of the source
- The `tee()` iterator function, which clones an iterator into several copies that can each be used independently

In this chapter, we'll look at the itertools functions that work with permutations and combinations. These include several functions and a few recipes built on these functions. The functions are as follows:

- `product()`: This function forms a Cartesian product equivalent to the nested `for` loops
- `permutations()`: This function emits tuples of length r from a universe p in all possible orderings; there are no repeated elements
- `combinations()`: This function emits tuples of length r from a universe p in sorted order; there are no repeated elements
- `combinations_with_replacement()`: This function emits tuples of length r from p in a sorted order, with repeated elements

These functions embody algorithms that iterate over potentially large result sets from small collections of input data. Some kinds of problems have exact solutions based on exhaustively enumerating a potentially gigantic universe of permutations. The functions make it simple to emit a large number of permutations; in some cases, the simplicity isn't actually optimal.

Enumerating the Cartesian product

The term **Cartesian product** refers to the idea of enumerating all the possible combinations of elements drawn from a number of sets.

Mathematically, we might say that the product of two sets, $\{1, 2, 3, \ldots, 13\} \times \{C, D, H, S\}$, has 52 pairs, as follows:

```
{(1, C), (1, D), (1, H), (1, S),
 (2, C), (2, D), (2, H), (2, S),
...,
 (13, C), (13, D), (13, H), (13, S)}
```

We can produce the preceding results by executing the following commands:

```
>>> list(product(range(1, 14), '♣♦♥♠'))
[(1, '♣'), (1, '♦'), (1, '♥'), (1, '♠'),
 (2, '♣'), (2, '♦'), (2, '♥'), (2, '♠'),
...
 (13, '♣'), (13, '♦'), (13, '♥'), (13, '♠')]
```

The calculation of a product can be extended to any number of iterable collections. Using a large number of collections can lead to a very large result set.

Reducing a product

In relational database theory, a join between tables can be thought of as a filtered product. A SQL `SELECT` statement that joins tables without a `WHERE` clause will produce a Cartesian product of rows in the tables. This can be thought of as the worst-case algorithm—a product without any filtering to pick the proper results. We can implement this using the itertools `product()` function to enumerate all possible combinations and filter those to keep the few that match properly.

We can define a `join()` function to join two iterable collections or generators, as shown in the following commands:

```
JT_ = TypeVar("JT_")
def join(
        t1: Iterable[JT_],
        t2: Iterable[JT_],
        where: Callable[[Tuple[JT_, JT_]], bool]
    ) -> Iterable[Tuple[JT_, JT_]]:
    return filter(where, product(t1, t2))
```

All combinations of the two iterables, `t1` and `t2`, are computed. The `filter()` function will apply the given `where()` function to pass or reject two-tuples, hinted as `Tuple[JT_, JT_]`, that match properly. The `where()` function has the hint `Callable[[Tuple[JT_, JT_]], bool]` to show that it returns a Boolean result. This is typical of how SQL database queries work in the worst-case situation where there are no useful indexes or cardinality statistics to suggest a better algorithm.

While this algorithm always works, it is terribly inefficient. We often need to look carefully at the problem and the available data to find a more efficient algorithm.

First, we'll generalize the problem slightly by replacing the simple Boolean matching function. Instead of a binary result, it's common to look for a minimum or maximum of some distance between items. In this case, the comparison will yield a float value.

Assume that we have a table of `Color` objects, as follows:

```
from typing import NamedTuple
class Color(NamedTuple):
    rgb: Tuple[int, int, int]
    name: str

[Color(rgb=(239, 222, 205), name='Almond'),
 Color(rgb=(255, 255, 153), name='Canary'),
 Color(rgb=(28, 172, 120), name='Green'),...
 Color(rgb=(255, 174, 66), name='Yellow Orange')]
```

For more information, see `Chapter 6`, *Recursions and Reductions*, where we showed you how to parse a file of colors to create `NamedTuple` objects. In this case, we've left the RGB as a `Tuple[int, int, int]`, instead of decomposing each individual field.

An image will have a collection of pixels:

```
pixels = [(r, g, b), (r, g, b), (r, g, b), ...]
```

As a practical matter, the **Python Imaging Library** (**PIL**) package presents the pixels in a number of forms. One of these is the mapping from the (x, y) coordinate to the RGB triple. For the Pillow project documentation, visit `https://pypi.python.org/pypi/Pillow`.

Given a `PIL.Image` object, we can iterate over the collection of pixels with something like the following commands:

```
from PIL import Image
from typing import Iterator, Tuple
Point = Tuple[int, int]
RGB = Tuple[int, int, int]
Pixel = Tuple[Point, RGB]
def pixel_iter(img: Image) -> Iterator[Pixel]:
    w, h = img.size
    return (
        (c, img.getpixel(c))
        for c in product(range(w), range(h))
    )
```

We've determined the range of each coordinate based on the image size, `img.size`. The calculation of the `product(range(w), range(h))` method creates all the possible combinations of coordinates. It is, effectively, two nested `for` loops.

This has the advantage of enumerating each pixel with its coordinates. We can then process the pixels in no particular order and still reconstruct an image. This is particularly handy when using multiprocessing or multithreading to spread the workload among several cores or processors. The `concurrent.futures` module provides an easy way to distribute work among cores or processors.

Computing distances

A number of decision-making problems require that we find a close enough match. We might not be able to use a simple equality test. Instead, we have to use a distance metric and locate items with the shortest distance to our target. For text, we might use the *Levenshtein distance*; this shows how many changes are required to get from a given block of text to our target.

We'll use a slightly simpler example. This will involve very simple math. However, even though it's simple, it doesn't work out well if we approach it naively.

When doing color matching, we won't have a simple equality test. We're rarely able to check for the exact equality of pixel colors. We're often forced to define a minimal distance function to determine whether two colors are close enough, without being the same three values of **R**, **G**, and **B**. There are several common approaches, including the Euclidean distance, Manhattan distance, and other complex weightings based on visual preferences.

Here are the Euclidean and Manhattan distance functions:

```
import math
def euclidean(pixel: RGB, color: Color) -> float:
    return math.sqrt(
        sum(map(
            lambda x, y: (x-y)**2,
            pixel,
            color.rgb)
        )
    )

def manhattan(pixel: RGB, color: Color) -> float:
    return sum(map(
        lambda x, y: abs(x-y),
        pixel,
        color.rgb)
    )
```

The **Euclidean distance** measures the hypotenuse of a right-angled triangle among the three points in an RGB space. The **Manhattan distance** sums the edges of each leg of the right-angled triangle among the three points. The Euclidean distance offers precision where the Manhattan distance offers calculation speed.

Looking forward, we're aiming for a structure that looks like this. For each individual pixel, we can compute the distance from that pixel's color to the available colors in a limited color set. The results of this calculation for a single pixel might look like this:

```
(
 ((0, 0), (92, 139, 195), Color(rgb=(239, 222, 205), name='Almond'),
169.10943202553784),
 ((0, 0), (92, 139, 195), Color(rgb=(255, 255, 153), name='Canary'),
204.42357985320578),
 ((0, 0), (92, 139, 195), Color(rgb=(28, 172, 120), name='Green'),
103.97114984456024),
 ((0, 0), (92, 139, 195), Color(rgb=(48, 186, 143), name='Mountain
Meadow'), 82.75868534480233),
 ((0, 0), (92, 139, 195), Color(rgb=(255, 73, 108), name='Radical Red'),
196.19887869200477),
 ((0, 0), (92, 139, 195), Color(rgb=(253, 94, 83),
```

```
 name='Sunset Orange'), 201.2212712413874),
 ((0, 0), (92, 139, 195), Color(rgb=(255, 174, 66), name='Yellow Orange'),
210.7961100210343)
)
```

We've shown an overall tuple that consists of a total of four tuples. Each of the four tuples contains the following contents:

- The pixel's coordinates; for example, `(0,0)`
- The pixel's original color; for example, `(92, 139, 195)`
- A `Color` object from our set of seven colors; for example, `Color(rgb=(239, 222, 205),name='Almond')`
- The Euclidean distance between the original color and the given `Color` object

We can see that the smallest Euclidean distance is a closest match color. This kind of reduction is done easily with the `min()` function. If the overall tuple is assigned to a variable name, `choices`, the pixel-level reduction would look like this:

```
min(choices, key=lambda xypcd: xypcd[3])
```

We've called each of the four tuples an `xypcd` value; that is, an *x-y* coordinate, pixel, color, and distance. The minimum distance calculation will then pick a single tuple as the optimal match between a pixel and color.

Getting all pixels and all colors

How do we get to the structure that contains all pixels and all colors? The answer is simple but, as we'll see, less than optimal.

One way to map pixels to colors is to enumerate all pixels and all colors using the `product()` function:

```
xy = lambda xyp_c: xyp_c[0][0]
p = lambda xyp_c: xyp_c[0][1]
c = lambda xyp_c: xyp_c[1]

distances = (
    (xy(item), p(item), c(item), euclidean(p(item), c(item)))
    for item in product(pixels, colors)
)
```

The core of this is the `product(pixel_iter(img), colors)` function that creates a sequence of all pixels combined with all colors. The overall expression then applies the `euclidean()` function to compute distances between pixels and `Color` objects. The result is a sequence of four tuples with the original *x-y* coordinate, the original pixel, an available color, and the distance between the original pixel color and the available color.

The final selection of colors uses the `groupby()` function and the `min(choices,...)` expression, as shown in the following command snippet:

```
for _, choices in groupby(
        distances, key=lambda xy_p_c_d: xy_p_c_d[0]):
    yield min(choices, key=lambda xypcd: xypcd[3])
```

The `product()` function applied to pixels and colors creates a long, flat iterable. We grouped the iterable into smaller collections where the coordinates match. This will break the big iterable into smaller iterables of just the pool of colors associated with a single pixel. We can then pick the minimal color distance for each available color for a pixel.

In a picture that's 3,648×2,736 with 133 Crayola colors, we have an iterable with 1,32,74,63,424 items to be evaluated. That is a billion combinations created by this `distances` expression. The number is not necessarily impractical. It's well within the limits of what Python can do. However, it reveals an important flaw in the naive use of the `product()` function.

We can't trivially do this kind of large-scale processing without first doing some analysis to see how large the intermediate data will be. Here are some `timeit` numbers for these two distance functions. This is the overall time to do each of these calculations only 1,000,000 times:

- Euclidean 2.8
- Manhattan 1.8

Scaling up by a factor of 1,000—from 1 million combinations to 1 billion—means the processing will take at least 1,800 seconds; that is, about half an hour for the Manhattan distance and 46 minutes to calculate the Euclidean distance. It appears this kind of naive bulk processing is ineffective for large datasets.

More importantly, we're doing it wrong. This kind of *width* × *height* × *color* processing is simply a bad design. In many cases, we can do much better.

Performance analysis

A key feature of any big data algorithm is locating a way to execute some kind of a divide-and-conquer strategy. This is true of functional programming design as well as imperative design.

We have three options to speed up this processing; they are as follows:

- We can try to use parallelism to do more of the calculations concurrently. On a four-core processor, the time can be cut to approximately 25 percent. This cuts the time to 8 minutes for Manhattan distances.
- We can see if caching intermediate results will reduce the amount of redundant calculation. The question arises of how many colors are the same and how many colors are unique.
- We can look for a radical change in the algorithm.

We'll combine the last two points by computing all the possible comparisons between source colors and target colors. In this case, as in many other contexts, we can easily enumerate the entire mapping and avoid redundant calculation when done on a pixel-by-pixel basis. We'll also change the algorithm from a series of comparisons to a series of simple lookups in a mapping object.

When looking at this idea of precomputing all transformations for source color to target color, we need some overall statistics for an arbitrary image. The code associated with this book includes `IMG_2705.jpg`. Here is a basic algorithm to collect all of the distinct color tuples from the specified image:

```
from collections import defaultdict, Counter
palette = defaultdict(list)
for xy, rgb in pixel_iter(img):
    palette[rgb].append(xy)

w, h = img.size
print("Total pixels", w*h)
print("Total colors", len(palette))
```

We collected all pixels of a given color into a list organized by color. From this, we'll learn the first of the following facts:

- The total number of pixels is 9,980,928. This fits the expectation for a 10 megapixel image.

- The total number of colors is 210,303. If we try to compute the Euclidean distance between actual colors and the 133 target colors, we would do 27,970,299 calculations, which might take about 76 seconds.
- Using a 3-bit mask, `0b11100000`, the total number of colors actually used is reduced to 214 out of a domain of $2^3 \times 2^3 \times 2^3 = 512$ possible colors.
- Using a 4-bit mask, `0b11110000`, 1,150 colors are actually used.
- Using a 5-bit mask, `0b11111000`, 5,845 colors are actually used.
- Using a 6-bit mask, `0b11111100`, 27,726 colors are actually used. The domain of possible colors swells to $2^6 \times 2^6 \times 2^6 = 262,144$.

This gives us some insight into how we can rearrange the data structure, calculate the matching colors quickly, and then rebuild the image without doing a billion comparisons.

The core idea behind masking is to preserve the most significant bits of a value and eliminate the least significant bits. Consider a color with a red value of 200. We can use the Python `bin()` function to see the binary representation of that value:

```
>>> bin(200)
'0b11001000'
>>> 200 & 0b11100000
192
>>> bin(192)
'0b11000000'
```

The computation of `200 & 0b11100000` applied a mask to conceal the least significant 5 bits and preserve the most significant 3 bits. What remains after the mask is applied as a red value of `192`.

We can apply mask values to the RGB three-tuple with the following command:

```
masked_color = tuple(map(lambda x: x&0b11100000, c))
```

This will pick out the most significant 3 bits of red, green, and blue values by using the `&` operator to select particular bits from an integer value. If we use this instead of the original color to create a `Counter` object, we'll see that the image only uses 214 distinct values after the mask is applied. This is fewer than half the theoretical number of colors.

Rearranging the problem

The naive use of the `product()` function to compare all pixels and all colors was a bad idea. There are 10 million pixels, but only 2,00,000 unique colors. When mapping the source colors to target colors, we only have to save 2,00,000 values in a simple map.

We'll approach it as follows:

1. Compute the source to target color mapping. In this case, let's use 3-bit color values as output. Each R, G, and B value comes from the eight values in the `range(0,256,32)` method. We can use this expression to enumerate all the output colors:

   ```
   product(range(0,256,32), range(0,256,32), range(0,256,32))
   ```

2. We can then compute the Euclidean distance to the nearest color in our source palette, doing just 68,096 calculations. This takes about 0.14 seconds. It's done one time only and computes the 2,00,000 mappings.
3. In one pass through the image, build a new image using the revised color table. In some cases, we can exploit the truncation of integer values. We can use an expression, such as `(0b11100000&r, 0b11100000&g, 0b11100000&b)` to remove the least significant bits of an image color. We'll look at this additional reduction in computation later.

This will replace a billion Euclidean distance calculations with 10 million dictionary lookups. This will replace 30 minutes of calculation with about 30 seconds of calculation.

Instead of doing color mapping for all pixels, we'll create a static mapping from input to output values. We can build the image building using simple lookup mapping from original color to a new color.

Once we have the palette of all 2,00,000 colors, we can apply the fast Manhattan distance to locate the nearest color in an output, such as the Crayola colors. This will use the algorithm for color matching shown earlier to compute the mapping instead of a result image. The difference will center on using the `palette.keys()` function instead of the `pixel_iter()` function.

We'll fold in yet another optimization—truncation. This will give us an even faster algorithm.

Combining two transformations

When combining multiple transformations, we can build a more complex mapping from the source through intermediate targets to the result. To illustrate this, we'll truncate the colors as well as apply a mapping.

In some problem contexts, truncation can be difficult. In other cases, it's often quite simple. For example, truncating US postal ZIP codes from nine to five characters is common. Postal codes can be further truncated to three characters to determine a regional facility that represents a larger geography.

For colors, we can use the bit-masking shown previously to truncate colors from three 8-bit values (24 bits, 16 million colors) to three 3-bit values (9 bits, 512 colors).

Here is a way to build a color map that combines both distances to a given set of colors and truncation of the source colors:

```
bit3 = range(0, 256, 0b100000)
best = (min((euclidean(rgb, c), rgb, c) for c in colors)
for rgb in product(bit3, bit3, bit3):
    color_map = dict(((b[1], b[2].rgb) for b in best))
```

We created a `range` object, `bit3`, that will iterate through all eight of the 3-bit color values. The use of the binary value, `0b100000`, can help visualize the way the bits are being used. The least significant 5 bits will be ignored; only the upper 3 bits will be used.

The `range` objects aren't like ordinary iterators; they can be used multiple times. As a result of this, the `product(bit3, bit3, bit3)` expression will produce all 512 color combinations that we'll use as the output colors.

For each truncated RGB color, we created a three-tuple that has (`0`) the distance from all crayon colors, (`1`) the RGB color, and (`2`) the crayon `Color` object. When we ask for the minimum value of this collection, we'll get the closest crayon `Color` object to the truncated RGB color.

We built a dictionary that maps from the truncated RGB color to the closest crayon. In order to use this mapping, we'll truncate a source color before looking up the nearest crayon in the mapping. This use of truncation coupled with the precomputed mapping shows how we might need to combine mapping techniques.

The following are the commands for the image replacement:

```
mask = 0b11100000
clone = img.copy()
for xy, rgb in pixel_iter(img):
    r, g, b = rgb
    repl = color_map[(mask&r, mask&g, mask&b)]
    clone.putpixel(xy, repl.rgb)
clone.show()
```

This uses the PIL `putpixel()` function to replace all of the pixels in a picture with other pixels. The mask value preserves the upper-most three bits of each color, reducing the number of colors to a subset.

What we've seen is that the naive use of some functional programming tools can lead to algorithms that are expressive and succinct, but also inefficient. The essential tools to compute the complexity of a calculation (sometimes called **Big-O analysis**) is just as important for functional programming as it is for imperative programming.

The problem is not that the `product()` function is inefficient. The problem is that we can use the `product()` function to create an inefficient algorithm.

Permuting a collection of values

When we permute a collection of values, we'll elaborate all the possible orders for the items. There are $n!$ ways to permute n items. We can use a sequence of permutations as a kind of brute-force solution to a variety of optimization problems.

By visiting `http://en.wikipedia.org/wiki/Combinatorial_optimization`, we can see that the exhaustive enumeration of all permutations isn't appropriate for larger problems. The use of the `itertools.permutations()` function is only handy for exploring very small problems.

One popular example of these combinatorial optimization problems is the assignment problem. We have *n* agents and *n* tasks, but the cost of each agent performing a given task is not equal. Imagine that some agents have trouble with some details, while other agents excel at these details. If we can properly assign tasks to agents, we can minimize the costs.

We can create a simple grid that shows how well a given agent is able to perform a given task. For a small problem of a half-dozen agents and tasks, there will be a grid of 36 costs. Each cell in the grid shows agents 0 to 5 performing tasks A to F.

We can easily enumerate all the possible permutations. However, this approach doesn't scale well. 10! is 36,28,800. We can see this sequence of 3 million items with the `list(permutations(range(10)))` method.

We would expect to solve a problem of this size in a few seconds. If we double the size of the problem to 20!, we would have a bit of a scalability problem: there would be 24,32,90,20,08,17,66,40,000 permutations. If it takes about 0.56 seconds to generate 10! permutations, so to generate 20! permutations, it would take about 12,000 years.

Assume that we have a cost matrix with 36 values that show the costs of six agents and six tasks. We can formulate the problem as follows:

```
perms = permutations(range(6))
alternatives = [
    (
        sum(
            cost[x][y] for y, x in enumerate(perm)
        ),
        perm
    )
    for perm in perms
]
m = min(alternatives)[0]
print([ans for s, ans in alternatives if s == m])
```

We've created all permutations of tasks for our six agents and assigned this to `perms`. From this, we've created two-tuples of the sum of all costs in the cost matrix and the permutation. To locate the relevant costs a specific permutation is enumerated to create two-tuples showing the agent and the assignment for that agent. For example, one of the permutations is (1, 3, 4, 2, 0, 5). The value of `list(enumerate((1, 3, 4, 2, 0, 5)))` is [(0, 1), (1, 3), (2, 4), (3, 2), (4, 0), (5, 5)], for a specific assignment of agents and tasks. The sum of the values in the `cost` matrix tells us how expensive this workplace setup would be.

The minimum cost is the optimal solution. In many cases, there might be multiple optimal solutions; this algorithm will locate all of them. The expression `min(alternatives)[0]` selects the first of the set of minima.

For small textbook examples, this is very fast. For larger examples, an approximation algorithm is more appropriate.

Generating all combinations

The `itertools` module also supports computing all combinations of a set of values. When looking at combinations, the order doesn't matter, so there are far fewer combinations than permutations. The number of combinations is often stated as $\binom{p}{r} = \frac{p!}{r!(p-r)!}$. This is the number of ways that we can take combinations of `r` things at a time from a universe of `p` items overall.

For example, there are 2,598,960 five-card poker hands. We can actually enumerate all 2 million hands by executing the following command:

```
hands = list(
    combinations(tuple(product(range(13), '♠♥♦♣')), 5))
```

More practically, assume we have a dataset with a number of variables. A common exploratory technique is to determine the correlation among all pairs of variables in a set of data. If there are v variables, then we will enumerate all variables that must be compared by executing the following command:

```
combinations(range(v), 2)
```

Let's get some sample data from `http://www.tylervigen.com` to show how this will work. We'll pick three datasets with the same time range: datasets numbered 7, 43, and 3,890. We'll simply laminate the data into a grid, repeating the year column.

This is how the first and the remaining rows of the yearly data will look:

```
[('year', 'Per capita consumption of cheese (US)Pounds (USDA)',
 'Number of people who died by becoming tangled in their
 bedsheetsDeaths (US) (CDC)',
 'year', 'Per capita consumption of mozzarella cheese (US)Pounds
 (USDA)', 'Civil engineering doctorates awarded (US)Degrees awarded
 (National Science Foundation)',
 'year', 'US crude oil imports from VenezuelaMillions of barrels
 (Dept. of Energy)', 'Per capita consumption of high fructose corn
 syrup (US)Pounds (USDA)'),
     (2000, 29.8, 327, 2000, 9.3, 480, 2000, 446, 62.6),
 (2001, 30.1, 456, 2001, 9.7, 501, 2001, 471, 62.5),
 (2002, 30.5, 509, 2002, 9.7, 540, 2002, 438, 62.8),
 (2003, 30.6, 497, 2003, 9.7, 552, 2003, 436, 60.9),
 (2004, 31.3, 596, 2004, 9.9, 547, 2004, 473, 59.8),
 (2005, 31.7, 573, 2005, 10.2, 622, 2005, 449, 59.1),
 (2006, 32.6, 661, 2006, 10.5, 655, 2006, 416, 58.2),
 (2007, 33.1, 741, 2007, 11, 701, 2007, 420, 56.1),
 (2008, 32.7, 809, 2008, 10.6, 712, 2008, 381, 53),
 (2009, 32.8, 717, 2009, 10.6, 708, 2009, 352, 50.1)]
```

This is how we can use the `combinations()` function to emit all the combinations of the nine variables in this dataset, taken two at a time:

```
combinations(range(9), 2)
```

There are 36 possible combinations. We'll have to reject the combinations that involve matching columns `year` and `year`. These will trivially correlate with a value of 1.00.

Here is a function that picks a column of data out of our dataset:

```
from typing import TypeVar, Iterator, Iterable
T_ = TypeVar("T_")
def column(source: Iterable[List[T_]], x: int) -> Iterator[T_]:
    for row in source:
        yield row[x]
```

This allows us to use the `corr()` function from `Chapter 4`, *Working with Collections*, to compare two columns of data.

This is how we can compute all combinations of correlations:

```
from itertools import *
from Chapter_4.ch04_ex4 import corr
for p, q in combinations(range(9), 2):
    header_p, *data_p = list(column(source, p))
    header_q, *data_q = list(column(source, q))
    if header_p == header_q:
        continue
    r_pq = corr(data_p, data_q)
    print("{2: 4.2f}: {0} vs {1}".format(header_p, header_q, r_pq))
```

For each combination of columns, we've extracted the two columns of data from our dataset. The `header_p, *data_p =` statement uses multiple assignments to separate the first item in the sequence, the header, from the remaining rows of data. If the headers match, we're comparing a variable to itself. This will be `True` for the three combinations of `year` and `year` that stem from the redundant year columns.

Given a combination of columns, we will compute the correlation function and then print the two headings along with the correlation of the columns. We've intentionally chosen some datasets that show spurious correlations with a dataset that doesn't follow the same pattern. In spite of this, the correlations are remarkably high.

The results look like this:

```
0.96: year vs Per capita consumption of cheese (US)Pounds (USDA)
0.95: year vs Number of people who died by becoming tangled in their
```

```
bedsheetsDeaths (US) (CDC)
0.92: year vs Per capita consumption of mozzarella cheese (US)Pounds (USDA)
0.98: year vs Civil engineering doctorates awarded (US) Degrees awarded
(National Science Foundation)
-0.80: year vs US crude oil imports from Venezuela Millions of barrels
(Dept. of Energy)
-0.95: year vs Per capita consumption of high fructose corn syrup (US)
Pounds (USDA)
0.95: Per capita consumption of cheese (US) Pounds (USDA) vs Number of
people who died by becoming tangled in their bedsheets Deaths (US) (CDC)
0.96: Per capita consumption of cheese (US)Pounds (USDA) vs year
0.98: Per capita consumption of cheese (US)Pounds (USDA) vs Per capita
consumption of mozzarella cheese (US)Pounds (USDA)
...
0.88: US crude oil imports from VenezuelaMillions of barrels (Dept. of
Energy) vs Per capita consumption of high fructose corn syrup (US)Pounds
(USDA)
```

It's not at all clear what this pattern means. Why do these values correlate? The presence of spurious correlations with no significance can cloud statistical analysis. We've located data that have strangely high correlations with no obvious causal factors.

What's important is that a simple expression, `combinations(range(9), 2)`, enumerates all the possible combinations of data. This kind of succinct, expressive technique makes it easier to focus on the data analysis issues instead of the Combinatoric algorithm considerations.

Recipes

The itertools chapter of the Python library documentation is outstanding. The basic definitions are followed by a series of recipes that are extremely clear and helpful. Since there's no reason to reproduce these, we'll reference them here. They are the required reading materials on functional programming in Python.

Section *10.1.2, Itertools Recipes,* in the *Python Standard Library* is a wonderful resource. Visit `https://docs.python.org/3/library/itertools.html#itertools-recipes` for more details.

These function definitions aren't importable functions in the itertools modules. These are ideas that need to be read and understood and then, perhaps, copied or modified before inclusion in an application.

The following table summarizes some recipes that show functional programming algorithms built from the itertools basics:

Function name	Arguments	Results
`powerset`	`(iterable)`	This generates all the subsets of the iterable. Each subset is actually a `tuple` object, not a set instance.
`random_product`	`(*args, repeat=1)`	This randomly selects from `itertools.product(*args, **kwds)`.
`random_permutation`	`(iterable, r=None)`	This randomly selects from `itertools.permutations(iterable, r)`.
`random_combination`	`(iterable, r)`	This randomly selects from `itertools.combinations(iterable, r)`.

Summary

In this chapter, we looked at a number of functions in the `itertools` module. This library module provides a number of functions that help us work with iterators in sophisticated ways.

We looked at the `product()` function that will compute all the possible combinations of the elements chosen from two or more collections. The `permutations()` function gives us different ways to reorder a given set of values. The `combinations()` function returns all the possible subsets of the original set.

We also looked at ways in which the `product()` and `permutations()` functions can be used naively to create extremely large result sets. This is an important cautionary note. A succinct and expressive algorithm can also involve a vast amount of computation. We must perform basic complexity analysis to be sure that the code will finish in a reasonable amount of time.

In the next chapter, we'll look at the `functools` module. This module includes some tools to work with functions as first-class objects. This builds on some material shown in `Chapter 2`, *Introducing Essential Functional Concepts*, and `Chapter 5`, *Higher-Order Functions*.

10
The Functools Module

Functional programming emphasizes functions as first-class objects. We've seen several higher-order functions that accept functions as arguments or return functions as results. In this chapter, we'll look at the `functools` library with some tools to help us implement some common functional design patterns.

We'll look at some higher-order functions. This extends the material from `Chapter 5`, *Higher-Order Functions*. We'll continue looking at higher-order function techniques in `Chapter 11`, *Decorator Design Techniques*, as well.

We'll look at the following functions in this module:

- `@lru_cache`: This decorator can be a huge performance boost for certain types of applications.
- `@total_ordering`: This decorator can help create rich comparison operators. Additionally, it lets us look at the more general question of object-oriented design mixed with functional programming.
- `partial()`: This function creates a new function from a function and some parameter value bindings.
- `reduce()`: This is a higher-order function that generalizes reductions such as `sum()`.

We'll defer two additional members of this library to `Chapter 11`, *Decorator Design Techniques*—the `update_wrapper()` and `wraps()` functions. We'll also look more closely at writing our own decorators in the next chapter.

We'll ignore the `cmp_to_key()` function entirely. Its purpose is to help with converting Python 2 code, which uses a comparison object, to run under Python 3, which uses key extraction. We're only interested in Python 3; we'll write proper key functions.

Function tools

We looked at a number of higher-order functions in Chapter 5, *Higher-Order Functions*. Those functions either accepted a function as an argument or returned a function (or generator expression) as a result. All those higher-order functions had an essential algorithm that was customized by injecting another function. Functions such as `max()`, `min()`, and `sorted()` accepted a `key=` function that customized their behavior. Functions such as `map()` and `filter()` accept a function and an iterable and apply this function to the arguments. In the case of the `map()` function, the results of the function are simply yielded. In the case of the `filter()` function, the Boolean result of the function is used to pass or reject values from an iterable source.

All the functions in Chapter 5, *Higher-Order Functions*, are part of the Python `__builtins__` package, they're available without the need to do an `import`. They are ubiquitous because they are so universally useful. The functions in this chapter must be introduced with an `import` because they're not quite so universally usable.

The `reduce()` function straddles this fence. It was originally built in. After some discussion, it was removed from the `__builtins__` package because of the possibility of abuse. Some seemingly simple operations can perform remarkably poorly.

Memoizing previous results with lru_cache

The `lru_cache` decorator transforms a given function into a function that might perform more quickly. The **LRU** means **Least Recently Used**—a finite pool of recently used items is retained. Items not frequently used are discarded to keep the pool to a bounded size.

Since this is a decorator, we can apply it to any function that might benefit from caching previous results. We can use it as follows:

```
from functools import lru_cache
@lru_cache(128)
def fibc(n: int) -> int:
    if n == 0: return 0
    if n == 1: return 1
    return fibc(n-1) + fibc(n-2)
```

This is an example based on `Chapter 6`, *Recursions and Reductions*. We've applied the `@lru_cache` decorator to the naive Fibonacci number calculation. Because of this decoration, each call to the `fibc(n)` function will now be checked against a cache maintained by the decorator. If the argument `n` is in the cache, the previously computed result is used instead of doing a potentially expensive re-calculation. Each return value is added to the cache.

We highlight this example because the naive recursion is quite expensive in this case. The complexity of computing any given Fibonacci number, F_n, involves not merely computing F_{n-1} but also F_{n-2}. This tree of values leads to a complexity in the order of $\mathbf{O}(2^n)$.

The argument value `128` is the size of the cache. This is used to limit the amount of memory used for the cache. When the cache is full, the LRU item is replaced.

We can try to confirm the benefits empirically using the `timeit` module. We can execute the two implementations a thousand times each to see how the timing compares. Using the `fib(20)` and `fibc(20)` methods shows just how costly this calculation is without the benefit of caching. Because the naive version is so slow, the `timeit` number of repetitions was reduced to only 1,000. Here are the results:

- Naive 3.23
- Cached 0.0779

Note that we can't trivially use the `timeit` module on the `fibc()` function. The cached values will remain in place, we'll only compute the `fibc(20)` function once, which populates this value in the cache. Each of the remaining 999 iterations will simply fetch the value from the cache. We need to actually clear the cache between uses of the `fibc()` function or the time drops to almost zero. This is done with a `fibc.cache_clear()` method built by the decorator.

The concept of memoization is powerful. There are many algorithms that can benefit from memoization of results.

The number of combinations of `p` things taken in groups of `r` is often stated as follows:

$$\binom{p}{r} = \frac{p!}{r!(p-r)!}$$

This binomial function involves computing three factorial values. It might make sense to use an `@lru_cache` decorator on a factorial function. A program that calculates a number of binomial values will not need to re-compute all of those factorials. For cases where similar values are computed repeatedly, the speedup can be impressive. For situations where the cached values are rarely reused, the overheads of maintaining the cached values outweigh any speedups.

When computing similar values repeatedly, we see the following:

- Naive factorial 0.174
- Cached factorial 0.046

It's important to recognize that the cache is a stateful object. This design pushes the edge of the envelope on purely functional programming. One functional ideal is to avoid changes of state. This concept of avoiding stateful variables is exemplified by a recursive function, the current state is contained in the argument values, and not in the changing values of variables. We've seen how tail-call optimization is an essential performance improvement to assure that this idealized recursion actually works nicely with the available processor hardware and limited memory budgets. In Python, we do this tail-call optimization manually by replacing the tail recursions with a `for` loop. Caching is a similar kind of optimization, we'll implement it manually as needed, knowing that it isn't purely functional programming.

In principle, each call to a function with an LRU cache has two results, the expected result and a new cache object available for future evaluations of the function. Pragmatically, the cache object is encapsulated inside the decorated version of the `fibc()` function, and it isn't available for inspection or manipulation.

Caching is not a panacea. Applications that work with float values might not benefit much from memoization because float values are often approximations. The least-significant bits of a float value should be seen as random noise that can prevent the exact equality test in the `lru_cache` decorator from working.

We'll revisit this in `Chapter 16`, *Optimizations and Improvements*. We'll look at some additional ways to implement this.

Defining classes with total ordering

The `total_ordering` decorator is helpful for creating new class definitions that implement a rich set of comparison operators. This might apply to numeric classes that subclass `numbers.Number`. It may also apply to semi-numeric classes.

As an example of a semi-numeric class, consider a playing card. It has a numeric rank and a symbolic suit. The rank matters only when doing simulations of some games. This is particularly important when simulating casino **blackjack**. Like numbers, cards have an ordering. We often sum the point values of each card, making them number-like. However, multiplication of *card* × *card* doesn't really make any sense; a card isn't quite like a number.

We can almost emulate a playing card with a `NamedTuple` base class as follows:

```
from typing import NamedTuple
class Card1(NamedTuple):
    rank: int
    suit: str
```

This is almost a good emulation. It suffers from a profound limitation: all comparisons include both the rank and the suit by default. This leads to the following awkward behavior when we compare the two of spades against the two of clubs:

```
>>> c2s= Card1(2, '\u2660')
>>> c2h= Card1(2, '\u2665')
>>> c2s
Card1(rank=2, suit='♣')
>>> c2h= Card1(2, '\u2665')
>>> c2h
Card1(rank=2, suit='♥')
>>> c2h == c2s
False
```

The default comparison rule doesn't work well for blackjack. It's also unsuitable for certain poker simulations.

For these games, it's better for the default comparisons between cards to be based on only their rank. Following is a more useful class definition:

```
from functools import total_ordering
from numbers import Number
from typing import NamedTuple

@total_ordering
class Card2(NamedTuple):
    rank: int
    suit: str
    def __eq__(self, other: Any) -> bool:
        if isinstance(other, Card2):
            return self.rank == other.rank
        elif isinstance(other, int):
            return self.rank == other
```

```
            return NotImplemented
        def __lt__(self, other: Any) -> bool:
            if isinstance(other, Card2):
                return self.rank < other.rank
            elif isinstance(other, int):
                return self.rank < other
            return NotImplemented
```

This class extends the `NamedTuple` class. We've provided a `__str__()` method to print a string representation of a `Card2` object.

There are two comparisons defined—one for equality and one for ordering. A wide variety of comparisons can be defined, and the `@total_ordering` decorator handles the construction of the remaining comparisons. In this case, the decorator created `__le__()`, `__gt__()`, and `__ge__()` from these two definitions. The default implementation of `__ne__()` uses `__eq__()`; this works without using a decorator.

Both of the methods provided allow two kinds of comparisons—between `Card2` objects, and also between a `Card2` object and an integer. The type hint must be `Any` to remain compatible with the superclass definition of `__eq__()` and `__lt__()`. It's clear that it could be narrowed to `Union[Card2, int]`, but this conflicts with the definition inherited from the superclass.

First, this class offers, proper comparison of only the ranks, as follows:

```
>>> c2s= Card2(2, '\u2660')
>>> c2h= Card2(2, '\u2665')
>>> c2h == c2s
True
>>> c2h == 2
True
>>> 2 == c2h
True
```

We can use this class for a number of simulations with simplified syntax to compare ranks of cards. Furthermore, the decorator builds a rich set of comparison operators as follows:

```
>>> c2s= Card2(2, '\u2660')
>>> c3h= Card2(3, '\u2665')
>>> c4c= Card2(4, '\u2663')
>>> c2s <= c3h < c4c
True
>>> c3h >= c3h
True
>>> c3h > c2s
```

```
True
>>> c4c != c2s
True
```

We didn't need to write all of the comparison method functions; they were generated by the decorator. The decorator's creation of operators isn't perfect. In our case, we've asked for comparisons with integers as well as between `Card` instances. This reveals some problems.

Operations such as the `c4c > 3` and `3 < c4c` comparisons would raise `TypeError` exceptions because of the way the operators are resolved. This is a limitation on what the `total_ordering` decorator can do. The problem rarely shows up in practice, since this kind of mixed-class coercion is relatively uncommon. In the rare case it's required, then the operators would have to be fully defined, and the `@total_ordering` decorator wouldn't be used.

Object-oriented programming is not antithetical to functional programming. There is a realm in which the two techniques are complementary. Python's ability to create immutable objects works particularly well with functional programming techniques. We can easily avoid the complexities of stateful objects, but still benefit from encapsulation to keep related method functions together. It's particularly helpful to define class properties that involve complex calculations; this binds the calculations to the class definition, making the application easier to understand.

Defining number classes

In some cases, we might want to extend the numeric tower available in Python. A subclass of `numbers.Number` may simplify a functional program. We can, for example, isolate parts of a complex algorithm to the `Number` subclass definition, making other parts of the application simpler or clearer.

Python already offers a rich variety of numeric types. The built-in types of the `int` and `float` variables cover a wide variety of problem domains. When working with currency, the `decimal.Decimal` package handles this elegantly. In some cases, we might find the `fractions.Fraction` class to be more appropriate than the `float` variable.

When working with geographic data, for example, we might consider creating a subclass of `float` variables that introduce additional attributes for conversion between degrees of latitude (or longitude) and radians. The arithmetic operations in this subclass could be done $mod(2\pi)$ to simplify calculations that move across the equator or the Greenwich meridian.

Since the Python Numbers class is intended to be immutable, ordinary functional design can be applied to all of the various method functions. The exceptional, Python, in-place, special methods (for example, the __iadd__() function) can be simply ignored.

When working with subclasses of Number, we have a fairly extensive volume of design considerations, as follows:

- Equality testing and hash value calculation. The core features of a hash calculation for numbers is documented in the *9.1.2 Notes for type implementors* section of the Python Standard Library.
- The other comparison operators (often defined via the @total_ordering decorator).
- The arithmetic operators—+, -, *, /, //, %, and **. There are special methods for the forward operations as well as additional methods for reverse type-matching. Given an expression such as a-b, Python uses the type of a to attempt to locate an implementation of the __sub__() method function, effectively, the a.__sub__(b) method. If the class of the left-hand value, a in this case, doesn't have the method or returns the NotImplemented exception, then the right-hand value is examined to see if the b.__rsub__(a) method provides a result. There's an additional special case that applies when class b is a subclass of class a this allows the subclass to override the left-hand side operation choice.
- The bit-fiddling operators—&, |, ^, >>, <<, and ~. These might not make sense for floating-point values; omitting these special methods might be the best design.
- Some additional functions like round(), pow(), and divmod() are implemented by numeric special method names. These might be meaningful for this class of numbers.

Chapter 7, *Additional Tuple Techniques*, provides a detailed example of creating a new type of number. Visit the link for more details: https://www.packtpub.com/application-development/mastering-object-oriented-python.

As we noted previously, functional programming and object-oriented programming can be complementary. We can easily define classes that follow functional programming design patterns. Adding new kinds of numbers is one example of leveraging Python's object-oriented features to create more readable functional programs.

Applying partial arguments with partial()

The `partial()` function leads to something called a **partial application**. A partially applied function is a new function built from an old function and a subset of the required arguments. It is closely related to the concept of currying. Much of the theoretical background is not relevant here, since currying doesn't apply to the way Python functions are implemented. The concept, however, can lead us to some handy simplifications.

We can look at trivial examples as follows:

```
>>> exp2 = partial(pow, 2)
>>> exp2(12)
4096
>>> exp2(17)-1
131071
```

We've created the function `exp2(y)`, which is the `pow(2, y)` function. The `partial()` function bounds the first positional parameter to the `pow()` function. When we evaluate the newly created `exp2()` function, we get values computed from the argument bound by the `partial()` function, plus the additional argument provided to the `exp2()` function.

The bindings of positional parameters are handled in a strict left-to-right order. For functions that accept keyword parameters, these can also be provided when building the partially applied function.

We can also create this kind of partially applied function with a lambda form as follows:

```
exp2 = lambda y: pow(2, y)
```

Neither is clearly superior. Measuring performance shows that the `partial()` function is slightly faster than a lambda form in the following manner:

- Partial 0.37
- Lambda 0.42

This is 0.05 seconds over 1 million iterations, not a remarkable saving.

Since lambda forms have all the capabilities of the `partial()` function, it seems that we can safely set this function aside as not being profoundly useful. We'll return to it in `Chapter 14`, *The PyMonad Library*, and look at how we can accomplish this with currying as well.

Reducing sets of data with the reduce() function

The `sum()`, `len()`, `max()`, and `min()` functions are in a way all specializations of a more general algorithm expressed by the `reduce()` function. The `reduce()` function is a higher-order function that folds a function into each pair of items in an iterable.

A sequence object is given as follows:

```
d = [2, 4, 4, 4, 5, 5, 7, 9]
```

The function `reduce(lambda x, y: x+y, d)` will fold in + operators to the list as follows:

```
2+4+4+4+5+5+7+9
```

It can help to include `()` to show the effective left-to-right grouping as follows:

```
((((((2+4)+4)+4)+5)+5)+7)+9
```

Python's standard interpretation of expressions involves a left-to-right evaluation of operators. Consequently, a fold left isn't a change in meaning. Some functional programming languages offer a fold-right alternative. When used in conjunction with recursion, a compiler for another language can do a number of clever optimizations. This isn't available in Python: a reduction is always left to right.

We can also provide an initial value as follows:

```
reduce(lambda x, y: x+y**2, iterable, 0)
```

If we don't, the initial value from the sequence is used as the initialization. Providing an initial value is essential when there's a `map()` function as well as a `reduce()` function. The following is how the right answer is computed with an explicit `0` initializer:

```
0 + 2**2 + 4**2 + 4**2 + 4**2 + 5**2 + 5**2 + 7**2 + 9**2
```

If we omit the initialization of 0, the `reduce()` function uses the first item as an initial value. This value does not have the transformation function applied, which leads to the wrong answer. In effect, the `reduce()` without a proper initial value is computing this:

```
2 + 4**2 + 4**2 + 4**2 + 5**2 + 5**2 + 7**2 + 9**2
```

This kind of mistake is part of the reason why `reduce()` must be used carefully.

We can define a number of built-in reductions using the `reduce()` higher-order function as follows:

```
sum2 = lambda data: reduce(lambda x, y: x+y**2, data, 0)
sum = lambda data: reduce(lambda x, y: x+y, data, 0)
count = lambda data: reduce(lambda x, y: x+1, data, 0)
min = lambda data: reduce(lambda x, y: x if x < y else y, data)
max = lambda data: reduce(lambda x, y: x if x > y else y, data)
```

The `sum2()` reduction function is the sum of squares, useful for computing the standard deviation of a set of samples. This `sum()` reduction function mimics the built-in `sum()` function. The `count()` reduction function is similar to the `len()` function, but it can work on an iterable, whereas the `len()` function can only work on a materialized `collection` object.

The `min()` and `max()` functions mimic the built-in reductions. Because the first item of the iterable is used for initialization, these two functions will work properly. If we provided any initial value to these `reduce()` functions, we might incorrectly use a value that never occurred in the original iterable.

Combining map() and reduce()

We can see how to build higher-order functions around these simple definitions. We'll show a simplistic map-reduce function that combines the `map()` and `reduce()` functions as follows:

```
from typing import Callable, Iterable, Any
def map_reduce(
        map_fun: Callable,
        reduce_fun: Callable,
        source: Iterable) -> Any:
    return reduce(reduce_fun, map(map_fun, source))
```

We've created a composite function from the `map()` and `reduce()` functions that take three arguments, the mapping transformation, the reduction operation, and the iterable or sequence source of items to process.

In its most generic form, shown previously, it's difficult to make any more formal assertions about the data types involved. The map and reduce functions can be very complex transformations. In the following examples, we'll use a slightly narrower definition for `map_reduce()` that looks as follows:

```
from typing import Callable, Iterable, TypeVar
T_ = TypeVar("T_")
def map_reduce(
        map_fun: Callable[[T_], T_],
        reduce_fun: Callable[[T_, T_], T_],
        source: Iterable[T_]) -> T_:
    return reduce(reduce_fun, map(map_fun, source))
```

This definition introduces a number of constraints. First, the iterator produces some consistently typed data. We'll bind the type to the `T_` type variable. Second, the map function accepts one argument of the type bound to `T_` and produces a result of the same type. Third, the reduce function will accept two arguments of this type, and return a result of the same type. For a simple, and essentially numeric application, this use of a consistent type variable, `T_`, works out nicely.

It's more common, however, to need a still narrower definition of the `map_fun()` function. Using a type such as `Callable[[T1_], T2_]` would capture the essence of a transformation from a source type, `T1_`, to a possibly distinct result type, `T2_`. The `reduce_fun()` function would then be `Callable[[T2_, T2_], T2_]` because these tend to preserve the type of the data.

We can build a sum-squared reduction using the `map()` and `reduce()` functions separately as follows:

```
def sum2_mr(source: Iterable[float]) -> float:
    return map_reduce(
        lambda y: y**2, lambda x, y: x+y, source)
```

In this case, we've used the `lambda y: y**2` parameter as a mapping to square each value. The reduction is the `lambda x, y: x+y` parameter. We don't need to explicitly provide an initial value because the initial value will be the first item in the iterable after the `map()` function has squared it.

The `lambda x, y: x+y` parameter is the + operator. Python offers all of the arithmetic operators as short functions in the `operator` module. The following is how we can slightly simplify our map-reduce operation:

```
from operator import add
def sum2_mr2(source: Iterable[float]) -> float:
    return map_reduce(lambda y: y**2, add, iterable)
```

We've used the `operator.add` method to sum our values instead of the longer lambda form.

The following is how we can count values in an iterable:

```
def count_mr(source: Iterable[float]) -> float:
    return map_reduce(lambda y: 1, lambda x, y: x+y, source)
```

We've used the `lambda y: 1` parameter to map each value to the value 1. The count is then a `reduce()` function using a lambda or the `operator.add` function.

The general-purpose `reduce()` function allows us to create any species of reduction from a large dataset to a single value. There are some limitations, however, on what we should do with the `reduce()` function.

We must avoid executing commands such as the following:

```
reduce(operator.add, list_of_strings, "")
```

Yes, this works. Python will apply the add operator between items of a string. However, the `"".join(list_of_strings)` method is considerably more efficient. A little study with `timeit` reveals that using `reduce()` to combine strings is $O(n^2)$ complex and very, very slow. It's difficult to spot these situations without some careful study of how operators are implemented for complex data structures.

Using the reduce() and partial() functions

The `sum()` function can be defined using the `partial(reduce, operator.add)` function. This, too, gives us a hint as to how we can create other mappings and other reductions. We can define all of the commonly used reductions as partials instead of lambdas:

```
sum2 = partial(reduce, lambda x, y: x+y**2)
count = partial(reduce, lambda x, y: x+1)
```

We can now use these functions via the `sum2(some_data)` or the `count(some_iter)` method. As we noted previously, it's not clear how much benefit this has. It's an important technique because it may be possible for a particularly complex calculation to be simplified using partial functions like this.

Using the map() and reduce() functions to sanitize raw data

When doing data cleansing, we'll often introduce filters of various degrees of complexity to exclude invalid values. We may also include a mapping to sanitize values in the cases where a valid but improperly formatted value can be replaced with a valid but proper value.

We might produce the following output:

```
def comma_fix(data: str) -> float:
    try:
        return float(data)
    except ValueError:
        return float(data.replace(",", ""))

def clean_sum(
        cleaner: Callable[[str], float],
        data: Iterable[str]
    ) -> float:
    return reduce(operator.add, map(cleaner, data))
```

We've defined a simple mapping, the `comma_fix()` class, that will convert data from a nearly correct string format into a usable floating-point value. This will remove the comma character. Another common variation will remove dollar signs and convert to `decimal.Decimal`.

We've also defined a map-reduce operation that applies a given cleaner function, the `comma_fix()` class in this case, to the data before doing a `reduce()` function, using the `operator.add` method.

We can apply the previously described function as follows:

```
>>> d = ('1,196', '1,176', '1,269', '1,240', '1,307',
... '1,435', '1,601', '1,654', '1,803', '1,734')
>>> clean_sum(comma_fix, d)
14415.0
```

We've cleaned the data, by fixing the commas, as well as computed a sum. The syntax is very convenient for combining these two operations.

We have to be careful, however, of using the cleaning function more than once. If we're also going to compute a sum of squares, we really should not execute the following command:

```
comma_fix_squared = lambda x: comma_fix(x)**2
```

If we use the `clean_sum(comma_fix_squared, d)` method as part of computing a standard deviation, we'll do the comma-fixing operation twice on the data, once to compute the sum and once to compute the sum of squares. This is a poor design; caching the results with an `lru_cache` decorator can help. Materializing the sanitized intermediate values as a temporary `tuple` object is better.

Using the groupby() and reduce() functions

A common requirement is to summarize data after partitioning it into groups. We can use a `defaultdict(list)` method to partition data. We can then analyze each partition separately. In Chapter 4, *Working with Collections,* we looked at some ways to group and partition. In Chapter 8, *The Itertools Module,* we looked at others.

The following is some sample data that we need to analyze:

```
>>> data = [('4', 6.1), ('1', 4.0), ('2', 8.3), ('2', 6.5),
... ('1', 4.6), ('2', 6.8), ('3', 9.3), ('2', 7.8),
... ('2', 9.2), ('4', 5.6), ('3', 10.5), ('1', 5.8),
... ('4', 3.8), ('3', 8.1), ('3', 8.0), ('1', 6.9),
... ('3', 6.9), ('4', 6.2), ('1', 5.4), ('4', 5.8)]
```

We've got a sequence of raw data values with a key and a measurement for each key.

One way to produce usable groups from this data is to build a dictionary that maps a key to a list of members in this groups as follows:

```
from collections import defaultdict
from typing import (
    Iterable, Callable, Dict, List, TypeVar,
    Iterator, Tuple, cast)
D_ = TypeVar("D_")
K_ = TypeVar("K_")

def partition(
        source: Iterable[D_],
        key: Callable[[D_], K_] = lambda x: cast(K_, x)
```

```
    ) -> Iterable[Tuple[K_, Iterator[D_]]]:
    pd: Dict[K_, List[D_]] = defaultdict(list)
    for item in source:
        pd[key(item)].append(item)
    for k in sorted(pd):
        yield k, iter(pd[k])
```

This will separate each item in the iterable into a group based on the key. The iterable source of data is described using a type variable of `D_`, representing the type of each data item. The `key()` function is used to extract a key value from each item. This function produces an object of some type, `K_`, that is generally distinct from the original data item type, `D_`. When looking at the sample data, the type of each data item is a `tuple`. The keys are of type `str`. The callable function for extracting a key transforms a `tuple` into a `string`.

This `key()` value extracted from each data item is used to append each item to a list in the `pd` dictionary. The `defaultdict` object is defined as mapping each key, `K_`, to a list of the data items, `List[D_]`.

The resulting value of this function matches the results of the `itertools.groupby()` function. It's an iterable sequence of the `(group key, iterator)` tuples. They group key will be of the type produced by the key function. The iterator will provide a sequence of the original data items.

Following is the same feature defined with the `itertools.groupby()` function:

```
from itertools import groupby

def partition_s(
        source: Iterable[D_],
        key: Callable[[D_], K_] = lambda x: cast(K_, x)
    ) -> Iterable[Tuple[K_, Iterator[D_]]]:
    return groupby(sorted(source, key=key), key)
```

The important difference in the inputs to each function is that the `groupby()` function version requires data to be sorted by the key, whereas the `defaultdict` version doesn't require sorting. For very large sets of data, the sort can be expensive, measured in both time and storage.

Here's the core partitioning operation. This might be used prior to filtering out a group, or it might be used prior to computing statistics for each group:

```
>>> for key, group_iter in partition(data, key=lambda x:x[0] ):
... print(key, tuple(group_iter))
1 (('1', 4.0), ('1', 4.6), ('1', 5.8), ('1', 6.9), ('1', 5.4))
2 (('2', 8.3), ('2', 6.5), ('2', 6.8), ('2', 7.8), ('2', 9.2))
3 (('3', 9.3), ('3', 10.5), ('3', 8.1), ('3', 8.0), ('3', 6.9))
4 (('4', 6.1), ('4', 5.6), ('4', 3.8), ('4', 6.2), ('4', 5.8))
```

We can summarize the grouped data as follows:

```
mean = lambda seq: sum(seq)/len(seq)
var = lambda mean, seq: sum((x-mean)**2/mean for x in seq)

Item = Tuple[K_, float]
def summarize(
        key_iter: Tuple[K_, Iterable[Item]]
    ) -> Tuple[K_, float, float]:
    key, item_iter = key_iter
    values = tuple(v for k, v in item_iter)
    m = mean(values)
    return key, m, var(m, values)
```

The results of the `partition()` functions will be a sequence of `(key, iterator)` two-tuples. The `summarize()` function accepts the two-tuple and and decomposes it into the key and the iterator over the original data items. In this function, the data items are defined as `Item`, a key of some type, `K_`, and a numeric value that can be coerced to `float`. From each two-tuple in the `item_iter` iterator we want the value portion, and we use a generator expression to create a tuple of only the values.

We can also use the expression `map(snd, item_iter)` to pick the second item from each of the two-tuples. This requires a definition of `snd = lambda x: x[1]`. The name `snd` is a short form of second to pick the second item of a tuple.

We can use the following command to apply the `summarize()` function to each partition:

```
>>> partition1 = partition(data, key=lambda x: x[0])
>>> groups1 = map(summarize, partition1)
```

The alternative commands are as follows:

```
>>> partition2 = partition_s(data, key=lambda x: x[0])
>>> groups2 = map(summarize, partition2)
```

Both will provide us with summary values for each group. The resulting group statistics look as follows:

```
1 5.34 0.93
2 7.72 0.63
3 8.56 0.89
4 5.5 0.7
```

The variance can be used as part of a χ^2 test to determine if the null hypothesis holds for this data. The null hypothesis asserts that there's nothing to see: the variance in the data is essentially random. We can also compare the data between the four groups to see if the various means are consistent with the null hypothesis or if there is some statistically significant variation.

Summary

In this chapter, we've looked at a number of functions in the `functools` module. This library module provides a number of functions that help us create sophisticated functions and classes.

We've looked at the `@lru_cache` function as a way to boost certain types of applications with frequent re-calculations of the same values. This decorator is of tremendous value for certain kinds of functions that take the `integer` or the `string` argument values. It can reduce processing by simply implementing memoization.

We looked at the `@total_ordering` function as a decorator to help us build objects that support rich ordering comparisons. This is at the fringe of functional programming, but is very helpful when creating new kinds of numbers.

The `partial()` function creates a new function with the partial application of argument values. As an alternative, we can build a `lambda` with similar features. The use case for this is ambiguous.

We also looked at the `reduce()` function as a higher-order function. This generalizes reductions like the `sum()` function. We'll use this function in several examples in later chapters. This fits logically with the `filter()` and `map()` functions as an important higher-order function.

In the following chapters, we'll look at how we can build higher-order functions using decorators. These higher-order functions can lead to slightly simpler and clearer syntax. We can use decorators to define an isolated aspect that we need to incorporate into a number of other functions or classes.

11
Decorator Design Techniques

Python offers us many ways to create higher-order functions. In `Chapter 5`, *Higher-Order Functions,* we looked at two techniques: defining a function that accepts a function as an argument, and defining a subclass of `Callable`, which is either initialized with a function or called with a function as an argument.

One of the benefits of decorated functions is that we can create composite functions. These are single functions that embody functionality from several sources. A composite function, $f \circ g(x)$, can be somewhat more expressive of a complex algorithm than $f(g(x))$. It's often helpful to have a number of syntax alternatives for expressing complex processing.

In this chapter, we'll look at the following topics:

- Using a decorator to build a function based on another function
- The `wraps()` function in the `functools` module; this can help us build decorators
- The `update_wrapper()` function, which may be helpful

Decorators as higher-order functions

The core idea of a decorator is to transform some original function into another form. A decorator creates a kind of composite function based on the decorator and the original function being decorated.

A decorator function can be used in one of the two following ways:

- As a prefix that creates a new function with the same name as the base function as follows:

```
@decorator
def original_function():
    pass
```

- As an explicit operation that returns a new function, possibly with a new name:

```
def original_function():
    pass
original_function = decorator(original_function)
```

These are two different syntaxes for the same operation. The prefix notation has the advantages of being tidy and succinct. The prefix location is more visible to some readers. The suffix notation is explicit and slightly more flexible.

While the prefix notation is common, there is one reason for using the suffix notation: we may not want the resulting function to replace the original function. We mayt want to execute the following command that allows us to use both the decorated and the undecorated functions:

```
new_function = decorator(original_function)
```

This will build a new function, named `new_function()`, from the original function. Python functions are first-class objects. When using the `@decorator` syntax, the original function is no longer available for use.

A decorator is a function that accepts a function as an argument and returns a function as the result. This basic description is clearly a built-in feature of the language. The open question then is how do we update or adjust the internal code structure of a function?

The answer is we don't. Rather than messing about with the inside of the code, it's much cleaner to define a new function that wraps the original function. It's easier to process the argument values or the result and leave the original function's core processing alone.

We have two phases of higher-order functions involved in defining a decorator; they are as follows:

- At definition time, a decorator function applies a wrapper to a base function and returns the new, wrapped function. The decoration process can do some one-time-only evaluation as part of building the decorated function. Complex default values can be computed, for example.

- At evaluation time, the wrapping function can (and usually does) evaluate the base function. The wrapping function can pre-process the argument values or post-process the return value (or both). It's also possible that the wrapping function may avoid calling the base function. In the case of managing a cache, for example, the primary reason for wrapping is to avoid expensive calls to the base function.

Here's an example of a simple decorator:

```
from functools import wraps
from typing import Callable, Optional, Any, TypeVar, cast

FuncType = Callable[..., Any]
F = TypeVar('F', bound=FuncType)

def nullable(function: F) -> F:
    @wraps(function)
    def null_wrapper(arg: Optional[Any]) -> Optional[Any]:
        return None if arg is None else function(arg)
    return cast(F, null_wrapper)
```

We almost always want to use the `functools.wraps()` function to assure that the decorated function retains the attributes of the original function. Copying the `__name__`, and `__doc__` attributes, for example, assures that the resulting decorated function has the name and docstring of the original function.

The resulting composite function, defined as the `null_wrapper()` function in the definition of the decorator, is also a type of higher-order function that combines the original function, the `function()` callable object, in an expression that preserves the `None` values. Within the resulting `null_wrapper()` function, the original `function` callable object is not an explicit argument; it is a free variable that will get its value from the context in which the `null_wrapper()` function is defined.

The decorator function's return value is the newly minted function. It will be assigned to the original function's name. It's important that decorators only return functions and that they don't attempt to process data. Decorators use meta-programming: a code that creates a code. The resulting `null_wrapper()` function, however, will be used to process the real data.

Note that the type hints use a feature of a `TypeVar` to assure that the result of applying the decorator will be a an object that's a type of `Callable`. The type variable `F` is bound to the original function's type; the decorator's type hint claims that the resulting function should have the same type as the argument function. A very general decorator will apply to a wide variety of functions, requiring a type variable binding.

We can apply our `@nullable` decorator to create a composite function as follows:

```
@nullable
def nlog(x: Optional[float]) -> Optional[float]:
    return math.log(x)
```

This will create a function, `nlog()`, which is a null-aware version of the built-in `math.log()` function. The decoration process returned a version of the `null_wrapper()` function that invokes the original `nlog()`. This result will be named `nlog()`, and will have the composite behavior of the wrapping and the original wrapped function.

We can use this composite `nlog()` function as follows:

```
>>> some_data = [10, 100, None, 50, 60]
>>> scaled = map(nlog, some_data)
>>> list(scaled)
[2.302585092994046, 4.605170185988092, None, 3.912023005428146,
4.0943445622221]
```

We've applied the function to a collection of data values. The `None` value politely leads to a `None` result. There was no exception processing involved.

This type of example isn't really suitable for unit testing. We'll need to round the values for testing purposes. For this, we'll need a null-aware `round()` function too.

Here's how we can create a null-aware rounding function using decorator notation:

```
@nullable
def nround4(x: Optional[float]) -> Optional[float]:
    return round(x, 4)
```

This function is a partial application of the `round()` function, wrapped to be null-aware. In some respects, this is a relatively sophisticated bit of functional programming that's readily available to Python programmers.

The `typing` module makes it particularly easy to describe the types of null-aware function and null-aware result, using the `Optional` type definition. The definition `Optional[float]` means `Union[None, float]`; either a `None` object or a `float` object may be used.

We could also create the null-aware rounding function using the following code:

```
nround4 = nullable(lambda x: round(x, 4))
```

Note that we didn't use the decorator in front of a function definition. Instead, we applied the decorator to a function defined as a lambda form. This has the same effect as a decorator in front of a function definition.

We can use this `round4()` function to create a better test case for our `nlog()` function as follows:

```
>>> some_data = [10, 100, None, 50, 60]
>>> scaled = map(nlog, some_data)
>>> [nround4(v) for v in scaled]
[2.3026, 4.6052, None, 3.912, 4.0943]
```

This result will be independent of any platform considerations. It's very handy for doctest testing.

It can be challenging to apply type hints to lambda forms. The following code shows what is required:

```
nround4l: Callable[[Optional[float]], Optional[float]] = (
    nullable(lambda x: round(x, 4))
)
```

The variable `nround4l` is given a type hint of `Callable` with an argument list of `[Optional[float]]` and a return type of `Optional[float]`. The use of the `Callable` hint is appropriate only for positional arguments. In cases where there will be keyword arguments or other complexities, see `http://mypy.readthedocs.io/en/latest/kinds_of_types.html#extended-callable-types`.

The `@nullable` decorator makes an assumption that the decorated function is `unary`. We would need to revisit this design to create a more general-purpose null-aware decorator that works with arbitrary collections of arguments.

In `Chapter 14`, *The PyMonad Library*, we'll look at an alternative approach to this problem of tolerating the `None` values. The `PyMonad` library defines a `Maybe` class of objects, which may have a proper value or may be the `None` value.

Using the functools update_wrapper() functions

The `@wraps` decorator applies the `update_wrapper()` function to preserve a few attributes of a wrapped function. In general, this does everything we need by default. This function copies a specific list of attributes from the original function to the resulting function created by a decorator. What's the specific list of attributes? It's defined by a module global.

The `update_wrapper()` function relies on a module global variable to determine what attributes to preserve. The `WRAPPER_ASSIGNMENTS` variable defines the attributes that are copied by default. The default value is this list of attributes to copy:

```
('__module__', '__name__', '__qualname__', '__doc__',
 '__annotations__')
```

It's difficult to make meaningful modifications to this list. The internals of the `def` statement aren't open to simple modification or change. This detail is mostly interesting as a piece of reference information.

If we're going to create `callable` objects, then we may have a class that provides some additional attributes as part of the definition. This could lead to a situation where a decorator must copy these additional attributes from the original wrapped `callable` object to the wrapping function being created. However, it seems simpler to make these kinds of changes through object-oriented class design, rather than exploit tricky decorator techniques.

Cross-cutting concerns

One general principle behind decorators is to allow us to build a composite function from the decorator and the original function to which the decorator is applied. The idea is to have a library of common decorators that can provide implementations for common concerns.

We often call these *cross-cutting* concerns because they apply across several functions. These are the sorts of things that we would like to design once through a decorator and have them applied in relevant classes throughout an application or a framework.

Concerns that are often centralized as decorator definitions include the following:

- Logging
- Auditing
- Security
- Handling incomplete data

A `logging` decorator, for example, may write standardized messages to the application's log file. An audit decorator may write details surrounding a database update. A security decorator may check some runtime context to be sure that the login user has the necessary permissions.

Our example of a *null-aware* wrapper for a function is a cross-cutting concern. In this case, we'd like to have a number of functions handle the `None` values by returning the `None` values instead of raising an exception. In applications where data is incomplete, we may have a need to process rows in a simple, uniform way without having to write lots of distracting `if` statements to handle missing values.

Composite design

The common mathematical notation for a composite function looks as follows:

$$f \circ g(x) = f(g(x))$$

The idea is that we can define a new function, $f \circ g(x)$, that combines two other functions, $f(y)$ and $g(x)$.

Python's multiple-line definition of a composition function can be done through the following code:

```
@f_deco
def g(x):
    something
```

The resulting function can be essentially equivalent to $f \circ g(x)$. The `f_deco()` decorator must define and return the composite function by merging an internal definition of `f()` with the provided `g()`.

The implementation details show that Python actually provides a slightly more complex kind of composition. The structure of a wrapper makes it helpful to think of Python decorator composition as follows:

$$w \circ g(x) = w_\beta \circ g \circ w_\alpha(x) = w_\beta\Big(g\big(w_\alpha(x)\big)\Big)$$

A decorator applied to some application function, $g(x)$, will include a wrapper function, w, that has two parts. One portion of the wrapper, $w_\alpha(y)$, applies to the arguments of the wrapped function, $g(x)$, and the other portion, $w_\beta(z)$, applies to the result of the wrapped function.

Here's a more concrete idea, shown as a `something_wrapper()` decorator definition:

```
@wraps(argument_function)
def something_wrapper(*args, **kw):
    # The "before" part, w_α, applied to *args or **kw
    result = argument_function(*args, **kw)
    # The "after" part, w_β, applied to the result
    return after_result
```

This shows the places to inject additional processing before and after the original function. This emphasizes an important distinction between the abstract concept of functional composition and the Python implementation: it's possible that a decorator can create either $f\big(g(x)\big)$, or $g\big(f(x)\big)$, or a more complex $f_\beta\Big(g\big(f_\alpha(x)\big)\Big)$. The syntax of decoration doesn't fully describe which kind of composition will be created.

The real value of decorators stems from the way any Python statement can be used in the wrapping function. A decorator can use `if` or `for` statements to transform a function into something used conditionally or iteratively. In the next section, the examples will leverage the `try` statement to perform an operation with a standard recovery from bad data. There are many clever things that can be done within this general framework.

A great deal of functional programming follows the $f \circ g(x) = f\big(g(x)\big)$ design pattern. Defining a composite from two smaller functions isn't always helpful. In some cases, it can be more informative to keep the two functions separate. In other cases, however, we may want to create a composite function to summarize processing.

It's very easy to create composites of the common higher-order functions, such as `map()`, `filter()`, and `reduce()`. Because these functions are simple, a composite function is often easy to describe, and helps make the programming more expressive.

For example, an application may include `map(f, map(g, x))`. It may be more clear to create a composite function and use a `map(f_g, x)` expression to describe applying a composite to a collection. Using `f_g = lambda x: f(g(x))` can often help explain a complex application as a composition of simpler functions.

It's important to note that there's no real performance advantage to either technique. The `map()` function is lazy: with two `map()` functions, one item will be taken from `x`, processed by the `g()` function, and then processed by the `f()` function. With a single `map()` function, an item will be taken from `x` and then processed by the `f_g()` composite function.

In `Chapter 14`, *The PyMonad Library*, we'll look at an alternative approach to this problem of creating composite functions from individual curried functions.

Preprocessing bad data

One cross-cutting concern in some exploratory data analysis applications is how to handle numeric values that are missing or cannot be parsed. We often have a mixture of `float`, `int`, and `Decimal` currency values that we'd like to process with some consistency.

In other contexts, we have *not applicable* or *not available* data values that shouldn't interfere with the main thread of the calculation. It's often handy to allow the `Not Applicable` values to pass through an expression without raising an exception. We'll focus on three bad-data conversion functions: `bd_int()`, `bd_float()`, and `bd_decimal()`. The composite feature we're adding will be defined before the built-in conversion function.

Here's a simple bad-data decorator:

```
import decimal
from typing import Callable, Optional, Any, TypeVar, cast

FuncType = Callable[..., Any]
F = TypeVar('F', bound=FuncType)

def bad_data(function: F) -> F:
    @wraps(function)
    def wrap_bad_data(text: str, *args: Any, **kw: Any) -> Any:
        try:
            return function(text, *args, **kw)
        except (ValueError, decimal.InvalidOperation):
            cleaned = text.replace(",", "")
            return function(cleaned, *args, **kw)
     return cast(F, wrap_bad_data)
```

This function wraps a given conversion function, with the name `function`, to try a second conversion in the event the first conversion involved bad data. The second conversion will be done after "," characters are removed. This wrapper passes the `*args` and `**kw` parameters into the wrapped function. This assures that the wrapped functions can have additional argument values provided.

The type variable `F` is bound to the original defined for the `function` parameter. The decorator is defined to return a function with the same type, `F`. The use of `cast()` provides a hint to the **mypy** tool that the wrapper doesn't change the signature of the function being wrapped.

We can use this wrapper to create bad-data sensitive functions as follows:

```
bd_int = bad_data(int)
bd_float = bad_data(float)
bd_decimal = bad_data(Decimal)
```

This will create a suite of functions that can do conversions of good data as well as a limited amount of data cleansing to handle specific kinds of bad data.

It's difficult to write type hints for some of these kinds of callable objects. In particular, the `int()` function has optional keyword arguments, for which type hints are complex. See `http://mypy.readthedocs.io/en/latest/kinds_of_types.html?highlight=keyword#extended-callable-types` for guidance on creating complex type signatures for callable objects.

The following are some examples of using the `bd_int()` function:

```
>>> bd_int("13")
13
>>> bd_int("1,371")
1371
>>> bd_int("1,371", base=16)
4977
```

We've applied the `bd_int()` function to a string that converted neatly and a string with the specific type of punctuation that we'll tolerate. We've also shown that we can provide additional parameters to each of these conversion functions.

We may like to have a more flexible decorator. One feature that we may like to add is the ability to handle a variety of data scrubbing alternatives. Simple `,` removal isn't always what we need. We may also need to remove `$`, or ° symbols, too. We'll look at more sophisticated, parameterized decorators in the next section.

Adding a parameter to a decorator

A common requirement is to customize a decorator with additional parameters. Rather than simply creating a composite $f \circ g(x)$, we can do something a bit more complex. With parameterized decorators, we can create $(f(c) \circ g)(x)$. We've applied a parameter, *c*, as part of creating the wrapper, $f(c)$. This parameterized composite function, $f(c) \circ g$, can then be used with the actual data, *x*.

In Python syntax, we can write it as follows:

```
@deco(arg)
def func(x):
    something
```

There are two steps to this. The first step applies the parameter to an abstract decorator to create a concrete decorator. Then the concrete decorator, the parameterized `deco(arg)` function, is applied to the base function definition to create the decorated function.

The effect is as follows:

```
def func(x):
    return something(x)
concrete_deco = deco(arg)
func= concrete_deco(func)
```

We've done three things, and they are as follows:

1. Defined a function, `func()`.
2. Applied the abstract decorator, `deco()`, to its argument, `arg`, to create a concrete decorator, `concrete_deco()`.
3. Applied the concrete decorator, `concrete_deco()`, to the base function to create the decorated version of the function; in effect it's `deco(arg)(func)`.

A decorator with arguments involves indirect construction of the final function. We seem to have moved beyond merely higher-order functions into something even more abstract: higher-order functions that create higher-order functions.

We can expand our bad-data aware decorator to create a slightly more flexible conversion. We'll define a `@bad_char_remove` decorator that can accept parameters of characters to remove. The following is a parameterized decorator:

```
import decimal
def bad_char_remove(
```

```
        *char_list: str
    ) -> Callable[[F], F]:
    def cr_decorator(function: F) -> F:
        @wraps(function)
        def wrap_char_remove(text: str, *args, **kw):
            try:
                return function(text, *args, **kw)
            except (ValueError, decimal.InvalidOperation):
                cleaned = clean_list(text, char_list)
                return function(cleaned, *args, **kw)
        return cast(F, wrap_char_remove)
    return cr_decorator
```

A parameterized decorator has two internal function definitions:

- An abstract decorator, the `cr_decorator` function. This will have its free variable, `char_list`, bound to become a concrete decorator. This decorator is then returned so it can be applied to a function. When applied, it will return a function wrapped inside the `wrap_char_remove` function. This has type hints with a type variable, `F`, that claim the wrapped function's type will be preserved by the wrapping operation.
- The decorating wrapper, the `wrap_char_remove` function, will replace the original function with a wrapped version. Because of the `@wraps` decorator, the `__name__` (and other attributes) of the new function will be replaced with the name of the base function being wrapped.

The overall decorator, the `bad_char_remove()` function's job is to bind the parameter to the abstract decorator and return the concrete decorator. The type hint clarifies the return value is a `Callable` object which transforms a `Callable` function into another `Callable` function. The language rules will then apply the concrete decorator to the following function definition.

Here's the `clean_list()` function used to remove all characters in a given list:

```
from typing import Tuple
def clean_list(text: str, char_list: Tuple[str, ...]) -> str:
    if char_list:
        return clean_list(
            text.replace(char_list[0], ""), char_list[1:])
    return text
```

This is a recursion because the specification is very simple. It can be optimized into a loop.

We can use the `@bad_char_remove` decorator to create conversion functions as follows:

```
@bad_char_remove("$", ",")
def currency(text: str, **kw) -> Decimal:
    return Decimal(text, **kw)
```

We've used our decorator to wrap a `currency()` function. The essential feature of the `currency()` function is a reference to the `decimal.Decimal` constructor.

This `currency()` function will now handle some variant data formats:

```
>>> currency("13")
Decimal('13')
>>> currency("$3.14")
Decimal('3.14')
>>> currency("$1,701.00")
Decimal('1701.00')
```

We can now process input data using a relatively simple `map(currency, row)` method to convert source data from strings to usable `Decimal` values. The `try:/except:` error-handling has been isolated to a function that we've used to build a composite conversion function.

We can use a similar design to create null-tolerant functions. These functions would use a similar `try:/except:` wrapper, but would simply return the `None` values.

Implementing more complex decorators

To create more complex functions, Python allows the following kinds of commands:

```
@f_wrap
@g_wrap
def h(x):
    something
```

There's nothing in Python to stop us from stacking up decorators that modify the results of other decorators. This has a meaning somewhat like $f \circ g \circ h(x)$. However, the resulting name will be merely $h(x)$. Because of this potential confusion, we need to be cautious when creating functions that involve deeply nested decorators. If our intent is simply to handle some cross-cutting concerns, then each decorator should be designed to handle a separate concern without creating much confusion.

If, on the other hand, we're using a decoration to create a composite function, it may also be better to use the following kinds of definitions:

```
from typing import Callable

m1: Callable[[float], float] = lambda x: x-1
p2: Callable[[float], float] = lambda y: 2**y
mersenne: Callable[[float], float] = lambda x: m1(p2(x))
```

Each of the variables has a type hint that defines the associated function. The three functions, `m1`, `p2`, and `mersenne`, have type hints of `Callable[[float], float]`, stating that this function will accept a number that can be coerced to float and return a number. Using a type definition such as `F_float = Callable[[float], float]` will avoid the simplistic repetition of the type definition.

If the functions are larger than simple expressions, the `def` statement is highly recommended. The use of lambda objects like this is a rare situation.

While many things can be done with decoration, it's essential to ask if using a decorator creates clear, succinct, expressive programming. When working with cross-cutting concerns, the features of the decorator are often essentially distinct from the function being decorated. This can be a wonderful simplification. Adding logging, debugging, or security checks through decoration is a widely followed practice.

One important consequence of an overly complex design is the difficulty in providing appropriate type hints. When the type hints devolve to simply using `Callable` and `Any` the design may have become too difficult to reason about clearly.

Complex design considerations

In the case of our data cleanup, the simplistic removal of stray characters may not be sufficient. When working with the geolocation data, we may have a wide variety of input formats that include simple degrees (`37.549016197`), degrees and minutes (`37° 32.94097′`), and degrees-minutes-seconds (`37° 32′ 56.46″`). Of course, there can be even more subtle cleaning problems: some devices will create an output with the Unicode U+00BA character, º, instead of the similar-looking degree character, °, which is U+00B0.

For this reason, it is often necessary to provide a separate cleansing function that's bundled in with the conversion function. This function will handle the more sophisticated conversions required by inputs that are as wildly inconsistent in format as latitudes and longitudes.

How can we implement this? We have a number of choices. Simple higher-order functions are a good choice. A decorator, on the other hand, doesn't work out terribly well. We'll look at a decorator-based design to see some limitations to what makes sense in a decorator.

The requirements have two orthogonal design considerations, and they are as follows:

1. The output conversion (`int`, `float`, `Decimal`).
2. The input cleaning (clean stray characters, reformat coordinates).

Ideally, one of these aspects is an essential function that gets wrapped and the other aspect is something that's included through a wrapper. The choice of essence versus wrapper isn't clear. One of the reasons it isn't clear is that our previous examples are a bit more complex than a simple two-part composite.

Considering the previous examples, it appears that this should be seen as a three-part composite:

- The output conversion (`int`, `float`, `Decimal`)
- The input cleansing: either a simple replace or a more complex multiple-character replacement
- The function that first attempted the conversion, then did any cleansing as a response to an exception, and attempted the conversion again

The third part—attempting the conversion and retrying—is the actual wrapper that also forms a part of the composite function. As we noted previously, a wrapper contains a argument phase and an return-value value, which we've called $w_\alpha(x)$ and $w_\beta(x)$, respectively.

We want to use this wrapper to create a composite of two additional functions. We have two choices for the design. We could include the cleansing function as an argument to the decorator on the conversion as follows:

```
@cleanse_before(cleanser)
def conversion(text):
    something
```

This first design claims that the conversion function is central, and the cleansing is an ancillary detail that will modify the behavior but preserve the original intent of the conversion.

Or, we could include the conversion function as an argument to the decorator for a cleansing function as follows:

```
@then_convert(converter)
def cleanse(text):
    something
```

This second design claims that the cleansing is central and the conversion is an ancillary detail. This is a bit confusing because the cleansing type is generally `Callable[[str], str]` where the conversion's type of `Callable[[str], some other type]` is what is required for the overall wrapped function.

While both of these approaches can create a usable composite function, the first version has an important advantage: the type signature of the `conversion()` function is also the type signature of the resulting composite function. This highlights a general design pattern for decorators: the type of the function being decorated is the easiest to preserve.

When confronted with several choices for defining a composite function, it is important to preserve the type hints for the function being decorated.

Consequently, the `@cleanse_before(cleaner)` style decorator is preferred. The decorator looks as follows:

```
def cleanse_before(
        cleanse_function: Callable
    ) -> Callable[[F], F]:
    def abstract_decorator(converter: F) -> F:
        @wraps(converter)
        def cc_wrapper(text: str, *args, **kw) -> Any:
            try:
                return converter(text, *args, **kw)
            except (ValueError, decimal.InvalidOperation):
                cleaned = cleanse_function(text)
                return converter(cleaned, *args, **kw)
        return cast(F, cc_wrapper)
    return abstract_decorator
```

We've defined a three-layer decorator. At the heart is the `cc_wrapper()` function that applies the `converter()` function. If this fails, then it uses the given `cleanse_function()` function and then tries the `converter()` function again. The `cc_wrapper()` function is built around the `cleanse_function()` and `converter()` functions by the `abstract_decorator()` concrete decorator function. The concrete decorator has the `cleanse_funcion()` function as a free variable. The concrete decorator is created by the decorator interface, `cleanse_before()`, which is customized by the `cleanse_function()` function.

The type hints emphasize the role of the `@cleanse_before` decorator. It expects some `Callable` function, named `cleanse_function`, and it creates a function, shown as `Callable[[F], F]`, which will transform a function into a wrapped function. This is a helpful reminder of how parameterized decorators work.

We can now build a slightly more flexible cleanse and convert function, `to_int()`, as follows:

```
def drop_punct2(text: str) -> str:
    return text.replace(",", "").replace("$", "")

@cleanse_before(drop_punct)
def to_int(text: str, base: int = 10) -> int:
    return int(text, base)
```

The integer conversion is decorated with a cleansing function. In this case, the cleansing function removes `$` and `,` characters. The integer conversion is wrapped by this cleansing.

The `to_int()` function defined previously leverages the built-in `int()` function. An alternative definition that avoids the `def` statement would be the following:

```
to_int2 = cleanse_before(drop_punct)(int)
```

This uses `drop_punct()` to wrap the built-in `int()` conversion function. Using the **mypy** tool `reveal_type()` function shows that the type signature for `to_int()` matches the type signature for the built-in `int()`.

We can use this enhanced integer conversion as follows:

```
>>> to_int("1,701")
1701
>>> to_int("97")
97
```

The type hints for the underlying `int()` function have been rewritten (and simplified) for the decorated function, `to_int()`. This is a consequence of trying to use decorators to wrap built-in functions.

Because of the complexity of defining parameterized decorators, it appears that this is the edge of the envelope. The decorator model doesn't seem to be ideal for this kind of design. It seems like a definition of a composite function would be more clear than the machinery required to build decorators.

Generally, decorators work well when we have a number of relatively simple and fixed aspects that we want to include with a given function (or a class). Decorators are also important when these additional aspects can be looked at as an infrastructure or as support, and not something essential to the meaning of the application code.

For something that involves multiple orthogonal design aspects, we may want to result to a callable class definition with various kinds of plugin strategy objects. This might have a simpler class definition than the equivalent decorator. Another alternative to decorators is to look closely at creating higher-order functions. In some cases, partial functions with various combinations of parameters may be simpler than a decorator.

The typical examples for cross-cutting concerns include logging or security testing. These features can be considered as the kind of background processing that isn't specific to the problem domain. When we have processing that is as ubiquitous as the air that surrounds us, then a decorator might be an appropriate design technique.

Summary

In this chapter, we've looked at two kinds of decorators: the simple decorator with no arguments and parameterized decorators. We've seen how decorators involve an indirect composition between functions: the decorator wraps a function (defined inside the decorator) around another function.

Using the `functools.wraps()` decorator assures that our decorators will properly copy attributes from the function being wrapped. This should be a piece of every decorator we write.

In the next chapter, we'll look at the multiprocessing and multithreading techniques that are available to us. These packages become particularly helpful in a functional programming context. When we eliminate a complex shared state and design around non-strict processing, we can leverage parallelism to improve the performance.

12
The Multiprocessing and Threading Modules

When we eliminate a complex, shared state and design around non-strict processing, we can leverage parallelism to improve performance. In this chapter, we'll look at the multiprocessing and multithreading techniques that are available to us. Python library packages become particularly helpful when applied to algorithms that permit lazy evaluation.

The central idea here is to distribute a functional program across several threads within a process or across several processes. If we've created a sensible functional design, we can avoid complex interactions among application components; we have functions that accept argument values and produce results. This is an ideal structure for a process or a thread.

In this chapter, we'll focus on several topics:

- The general idea of functional programming and concurrency.
- What concurrency really means when we consider cores, CPUs, and OS-level parallelism. It's important to note that concurrency won't magically make a bad algorithm faster.
- Using built-in `multiprocessing` and `concurrent.futures` modules. These modules allow a number of parallel execution techniques. The `dask` package can do much of this, also.

We'll focus on process-level parallelism more than multithreading. Using process parallelism allows us to completely ignore Python's **Global Interpreter Lock** (**GIL**). For more information on Python's GIL, see `https://docs.python.org/3.3/c-api/init.html#thread-state-and-the-global-interpreter-lock`.

Functional programming and concurrency

The most effective concurrent processing occurs where there are no dependencies among the tasks being performed. The biggest difficulty in developing concurrent (or *parallel*) programming is coordinating updates to shared resources.

When following functional design patterns, we tend to avoid stateful programs. A functional design should minimize or eliminate concurrent updates to shared objects. If we can design software where lazy, non-strict evaluation is central, we can also design software where concurrent evaluation is helpful. This can lead to *embarrassingly parallel* design, where most of the work can be done concurrently with few or no interactions among computations.

Dependencies among operations are central to programming. In the `2*(3+a)` expression, the `(3+a)` subexpression must be evaluated first. The overall value of the expression depends on the proper ordering of the two operations.

When working with a collection, we often have situations where the processing order among items in the collection doesn't matter. Consider the following two examples:

```
x = list(func(item) for item in y)
x = list(reversed([func(item) for item in y[::-1]]))
```

Both of these commands have the same result even though the `func(item)` expressions are evaluated in opposite orders. This works when each evaluation of `func(item)` is independent and free of side effects.

Even the following command snippet has the same result:

```
import random
indices = list(range(len(y)))
random.shuffle(indices)
x = [None]*len(y)
for k in indices:
    x[k] = func(y[k])
```

The evaluation order in the preceding example is random. Because each evaluation `func(y[k])` is independent of all others, the order of evaluation doesn't matter. This is the case with many algorithms that permit non-strict evaluation.

What concurrency really means

In a small computer, with a single processor and a single core, all evaluations are serialized through the one and only core of the processor. The use OS throughout will interleave multiple processes and multiple threads through clever time-slicing arrangements.

On a computer with multiple CPUs or multiple cores in a single CPU, there can be some actual concurrent processing of CPU instructions. All other concurrency is simulated through time slicing at the OS level. A macOS X laptop can have 200 concurrent processes that share the CPU; this is many more processes than the number of available cores. From this, we can see that OS time slicing is responsible for most of the apparently concurrent behavior of the system as a whole.

The boundary conditions

Let's consider a hypothetical algorithm that has a complexity described by $\mathbf{O}(n^2)$. Assume that there is an inner loop that involves 1,000 bytes of Python code. When processing 10,000 objects, we're executing 100 billion Python operations. We can call this the essential processing budget. We can try to allocate as many processes and threads as we feel might be helpful, but the processing budget can't change.

An individual CPython bytecode doesn't have simple execution timing. However, a long-term average on a macOS X laptop shows that we can expect about 60 MB of code to be executed per second. This means that our 100 billion bytecode operation will take about 1,666 seconds, or 28 minutes.

If we have a dual-processor, four-core computer, then we might cut the elapsed time to 25% of the original total: about 7 minutes. This presumes that we can partition the work into four (or more) independent OS processes.

The important consideration here is that the overall budget of 100 billion bytecodes can't be changed. Parallelism won't magically reduce the workload. It can only change the schedule to perhaps reduce the elapsed time.

Switching to a better algorithm, $\mathbf{O}(n \log n)$, can reduce the workload from 100 billion operations to 133 million operations with a potential runtime of about 2 seconds. Spread over four cores, we might see a response in 516 milliseconds. Parallelism can't have the kind of dramatic improvements that algorithm change will have.

Sharing resources with process or threads

The OS ensures that there is little or no interaction between processes. When creating a multiprocessing application, where multiple processes must interact, a common OS resource must be explicitly shared. This can be a common file, a specific, shared-memory object, or a semaphore with a shared state between the processes. Processes are inherently independent; interaction among them is exceptional.

Multiple threads, in contrast, are part of a single process; all threads of a process share OS resources. We can make an exception to get some thread-local memory that can be freely written without interference from other threads. Outside thread-local memory, operations that write to memory can set the internal state of the process in a potentially unpredictable order. Explicit locking must be used to avoid problems. As noted previously, the overall sequence of instruction executions is rarely, strictly speaking, concurrent. The instructions from concurrent threads and processes are generally interleaved among the cores in an unpredictable order. With threading comes the possibility of destructive updates to shared variables and the need for mutually exclusive access through locking.

At the *bare metal* hardware level, there are some complex memory write situations. For more information on issues with memory writes, visit `http://en.wikipedia.org/wiki/Memory_disambiguation`.

The existence of concurrent object updates can create havoc when trying to design multithreaded applications. Locking is one way to avoid concurrent writes to shared objects. Avoiding shared objects in general is another viable design technique. The second technique—avoiding writes to shared objects—is more applicable to functional programming.

In CPython, the GIL is used to ensure that OS thread scheduling will not interfere with the internals of maintaining Python data structures. In effect, the GIL changes the granularity of scheduling from machine instructions to groups of Python virtual machine operations.

The impact of the GIL in term of ensuring data structure integrity is often negligible. The greatest impact on performance comes from the choice of algorithm.

Where benefits will accrue

A program that does a great deal of calculation and relatively little I/O will not see much benefit from concurrent processing. If a calculation has a budget of 28 minutes of computation, then interleaving the operations in different ways won't have a dramatic impact. Using eight cores may cut the time by approximately one-eighth. The actual time savings depend on the OS and language overheads, which are difficult to predict. Introducing concurrency will not have the kind of performance impact that a better algorithm will have.

When a calculation involves a great deal of I/O, then interleaving CPU processing with I/O requests will *dramatically* improve performance. The idea is to do computations on some pieces of data while waiting for the OS to complete the I/O of other pieces of data. Because I/O generally involves a great deal of waiting, an eight-core processor can interleave the work from dozens of concurrent I/O requests.

We have two approaches to interleaving computation and I/O. They are as follows:

- We can create a pipeline of processing stages. An individual item must move through all of the stages where it is read, filtered, computed, aggregated, and written. The idea of multiple concurrent stages is to have distinct data objects in each stage. Time slicing among the stages will allow computation and I/O to be interleaved.
- We can create a pool of concurrent workers, each of which we perform all of the processing for a data item. The data items are assigned to workers in the pool and the results are collected from the workers.

The differences between these approaches aren't crisp; there is a blurry middle region that's clearly not one or the other. It's common to create a hybrid mixture where one stage of a pipeline involves a pool of workers to make that stage as fast as the other stages. There are some formalisms that make it somewhat easier to design concurrent programs. The **Communicating Sequential Processes** (**CSP**) paradigm can help design message-passing applications. Packages such as `pycsp` can be used to add CSP formalisms to Python.

I/O-intensive programs often gain the most dramatic benefits from concurrent processing. The idea is to interleave I/O and processing. CPU-intensive programs will see smaller benefits from concurrent processing.

Using multiprocessing pools and tasks

Concurrency is a form of *non-strict* evaluation: the exact order of operations is unpredictable. The `multiprocessing` package introduces the concept of a `Pool` object. A `Pool` object contains a number of worker processes and expects these processes to be executed concurrently. This package allows OS scheduling and time slicing to interleave execution of multiple processes. The intention is to keep the overall system as busy as possible.

To make the most of this capability, we need to decompose our application into components for which non-strict concurrent execution is beneficial. The overall application must be built from discrete tasks that can be processed in an indefinite order.

An application that gathers data from the internet through web scraping, for example, is often optimized through parallel processing. We can create a `Pool` object of several identical workers, which implement the website scraping. Each worker is assigned tasks in the form of URLs to be analyzed.

An application that analyzes multiple log files is also a good candidate for parallelization. We can create a `Pool` object of analytical workers. We can assign each log file to an analytical worker; this allows reading and analysis to proceed in parallel among the various workers in the `Pool` object. Each individual worker will be performing both I/O and computation. However, one worker can be analyzing while other workers are waiting for I/O to complete.

Because the benefits depend on difficult-to-predict timing for input and output operations, multiprocessing always involves experimentation. Changing the pool size and measuring elapsed time is an essential part of designing concurrent applications.

Processing many large files

Here is an example of a multiprocessing application. We'll scrape **Common Log Format** (**CLF**) lines in web log files. This is the generally used format for web server access logs. The lines tend to be long, but look like the following when wrapped to the book's margins:

```
99.49.32.197 - - [01/Jun/2012:22:17:54 -0400] "GET /favicon.ico
 HTTP/1.1" 200 894 "-" "Mozilla/5.0 (Windows NT 6.0)
 AppleWebKit/536.5 (KHTML, like Gecko) Chrome/19.0.1084.52
 Safari/536.5"
```

We often have large numbers of files that we'd like to analyze. The presence of many independent files means that concurrency will have some benefit for our scraping process.

We'll decompose the analysis into two broad areas of functionality. The first phase of processing is the essential parsing of the log files to gather the relevant pieces of information. We'll further decompose the parsing into four stages. They are as follows:

1. All the lines from multiple source log files are read.
2. Then, we create simple namedtuple objects from the lines of log entries in a collection of files.
3. The details of more complex fields such as dates and URLs are parsed.
4. Uninteresting paths from the logs are rejected; we can also think of this as parsing only the interesting paths.

Once past the parsing phase, we can perform a large number of analyses. For our purposes in demonstrating the `multiprocessing` module, we'll look at a simple analysis to count occurrences of specific paths.

The first portion, reading from source files, involves the most input processing. Python's use of file iterators will translate into lower-level OS requests for the buffering of data. Each OS request means that the process must wait for the data to become available.

Clearly, we want to interleave the other operations so that they are not waiting for I/O to complete. We can interleave operations along a spectrum from individual rows to whole files. We'll look at interleaving whole files first, as this is relatively simple to implement.

The functional design for parsing Apache CLF files can look as follows:

```
data = path_filter(
    access_detail_iter(
        access_iter(
            local_gzip(filename))))
```

We've decomposed the larger parsing problem into a number of functions that will handle each portion of the parsing problem. The `local_gzip()` function reads rows from locally cached GZIP files. The `access_iter()` function creates a `NamedTuple` object for each row in the access log. The `access_detail_iter()` function will expand on some of the more difficult-to-parse fields. Finally, the `path_filter()` function will discard some paths and file extensions that aren't of much analytical value.

It can help to visualize this kind of design as a pipeline of processing, as shown here:

```
(local_gzip(filename) | access_iter | access_detail_iter | path_filter) >
data
```

This uses shell notation of a pipe (|) to pass data from process to process. The built-in Python `pipes` module facilitates building actual shell pipelines to leverage OS multiprocessing capabilities. Other packages such as `pipetools` or `pipe` provide a similar way to visualize a composite function.

Parsing log files – gathering the rows

Here is the first stage in parsing a large number of files: reading each file and producing a simple sequence of lines. As the log files are saved in the `.gzip` format, we need to open each file with the `gzip.open()` function instead of the `io.open()` function or the `__builtins__.open()` function.

The `local_gzip()` function reads lines from locally cached files, as shown in the following command snippet:

```
from typing import Iterator
def local_gzip(pattern: str) -> Iterator[Iterator[str]]:
    zip_logs= glob.glob(pattern)
    for zip_file in zip_logs:
        with gzip.open(zip_file, "rb") as log:
            yield (
                line.decode('us-ascii').rstrip()
                for line in log)
```

The preceding function iterates through all files that match the given pattern. For each file, the yielded value is a generator function that will iterate through all lines within that file. We've encapsulated several things, including wildcard file matching, the details of opening a log file compressed with the `.gzip` format, and breaking a file into a sequence of lines without any trailing newline (`\n`) characters.

The essential design pattern here is to yield values that are generator expressions for each file. The preceding function can be restated as a function and a mapping that applies that particular function to each file. This can be useful in the rare case when individual files need to be identified. In some cases, this can be optimized to use `yield from` to make all of the various log files appear to be a single stream of lines.

There are several other ways to produce similar output. For example, here is an alternative version of the inner `for` loop in the preceding example. The `line_iter()` function will also emit lines of a given file:

```
def line_iter(zip_file: str) -> Iterator[str]:
    log = gzip.open(zip_file, "rb")
    return (line.decode('us-ascii').rstrip() for line in log)
```

The `line_iter()` function applies the `gzip.open()` function and some line cleanup. We can use mapping to apply the `line_iter()` function to all files that match a pattern, as follows:

```
map(line_iter, glob.glob(pattern))
```

While this alternative mapping is succinct, it has the disadvantage of leaving open file objects lying around waiting to be properly garbage-collected when there are no more references. When processing a large number of files, this seems like a needless bit of overhead. For this reason, we'll focus on the `local_gzip()` function shown previously.

The previous alternative mapping has the distinct advantage of fitting in well with the way the `multiprocessing` module works. We can create a worker pool and map tasks (such as file reading) to the pool of processes. If we do this, we can read these files in parallel; the open file objects will be part of separate processes.

An extension to this design will include a second function to transfer files from the web host using FTP. As the files are collected from the web server, they can be analyzed using the `local_gzip()` function.

The results of the `local_gzip()` function are used by the `access_iter()` function to create namedtuples for each row in the source file that describes file access.

Parsing log lines into namedtuples

Once we have access to all of the lines of each log file, we can extract details of the access that's described. We'll use a regular expression to decompose the line. From there, we can build a `namedtuple` object.

Here is a regular expression to parse lines in a CLF file:

```
import re
format_pat = re.compile(
    r"(?P<host>[\d\.]+)\s+"
    r"(?P<identity>\S+)\s+"
```

```
        r"(?P<user>\S+)\s+"
        r"\[(?P<time>.+?)\]\s+"
        r'"(?P<request>.+?)"\s+'
        r"(?P<status>\d+)\s+"
        r"(?P<bytes>\S+)\s+"
        r'"(?P<referer>.*?)"\s+' # [SIC]
        r'"(?P<user_agent>.+?)"\s*'
    )
```

We can use this regular expression to break each row into a dictionary of nine individual data elements. The use of `[]` and `"` to delimit complex fields such as the `time`, `request`, `referrer`, and `user_agent` parameters can be handled elegently by transforming the text into a `NamedTuple` object.

Each individual access can be summarized as a subclass of `NamedTuple`, as follows:

```
from typing import NamedTuple
class Access(NamedTuple):
    host: str
    identity: str
    user: str
    time: str
    request: str
    status: str
    bytes: str
    referer: str
    user_agent: str
```

We've taken pains to ensure that the `NamedTuple` field names match the regular expression group names in the `(?P<name>)` constructs for each portion of the record. By making sure the names match, we can very easily transform the parsed dictionary into a tuple for further processing.

Here is the `access_iter()` function that requires each file to be represented as an iterator over the lines of the file:

```
from typing import Iterator
def access_iter(
        source_iter: Iterator[Iterator[str]]
    ) -> Iterator[Access]:
    for log in source_iter:
        for line in log:
            match = format_pat.match(line)
            if match:
                yield Access(**match.groupdict())
```

The output from the `local_gzip()` function is a sequence of sequences. The outer sequence is based on the individual log files. For each file, there is a nested iterable sequence of lines. If the line matches the given pattern, it's a file access of some kind. We can create an `Access` namedtuple from the `match` dictionary. Non-matching lines are quietly discarded.

The essential design pattern here is to build an immutable object from the results of a parsing function. In this case, the parsing function is a regular expression matcher. Other kinds of parsing will fit this design pattern.

There are some alternative ways to do this. For example, we can use the `map()` function as follows:

```
def access_builder(line: str) -> Optional[Access]:
    match = format_pat.match(line)
    if match:
        return Access(**match.groupdict())
    return None
```

The preceding alternative function embodies just the essential parsing and construction of an `Access` object. It will either return an `Access` or a `None` object. Here is how we can use this function to flatten log files into a single stream of `Access` objects:

```
filter(
    None,
    map(
        access_builder,
        (line for log in source_iter for line in log)
    )
)
```

This shows how we can transform the output from the `local_gzip()` function into a sequence of `Access` instances. In this case, we apply the `access_builder()` function to the nested iterator of the iterable structure that results from reading a collection of files. The `filter()` function removes `None` objects from the result of the `map()` function.

Our point here is to show that we have a number of functional styles for parsing files. In `Chapter 4`, *Working with Collections*, we looked at very simple parsing. Here, we're performing more complex parsing, using a variety of techniques.

Parsing additional fields of an Access object

The initial `Access` object created previously doesn't decompose some inner elements in the nine fields that comprise an access log line. We'll parse those items separately from the overall decomposition into high-level fields. Doing these parsing operations separately makes each stage of processing simpler. It also allows us to replace one small part of the overall process without breaking the general approach to analyzing logs.

The resulting object from the next stage of parsing will be a `NamedTuple` subclass, `AccessDetails`, which wraps the original `Access` tuple. It will have some additional fields for the details parsed separately:

```
from typing import NamedTuple, Optional
import datetime
import urllib.parse

class AccessDetails(NamedTuple):
    access: Access
    time: datetime.datetime
    method: str
    url: urllib.parse.ParseResult
    protocol: str
    referrer: urllib.parse.ParseResult
    agent: Optional[AgentDetails]
```

The `access` attribute is the original `Access` object, a collection of simple strings. The `time` attribute is the parsed `access.time` string. The `method`, `url`, and `protocol` attributes come from decomposing the `access.request` field. The `referrer` attribute is a parsed URL.

The `agent` attribute can also be broken down into fine-grained fields. An unconventional browser or website scraper can produce an `agent` string that can't be parsed, so this attribute is marked with the `Optional` type hint.

Here are the attributes that comprise the `AgentDetails` subclass of the `NamedTuple` class:

```
class AgentDetails(NamedTuple):
    product: str
    system: str
    platform_details_extensions: str
```

These fields reflect the most common syntax for agent descriptions. There is considerable variation in this area, but this particular subset of values seems to be reasonably common.

Here are the three detail-level parsers for the fields to be decomposed:

```
from typing import Tuple, Optional
import datetime
import re

def parse_request(request: str) -> Tuple[str, str, str]:
    words = request.split()
    return words[0], ' '.join(words[1:-1]), words[-1]

def parse_time(ts: str) -> datetime.datetime:
    return datetime.datetime.strptime(
        ts, "%d/%b/%Y:%H:%M:%S %z"
    )

agent_pat = re.compile(
    r"(?P<product>\S*?)\s+"
    r"\((?P<system>.*?)\)\s*"
    r"(?P<platform_details_extensions>.*)"
)

def parse_agent(user_agent: str) -> Optional[AgentDetails]:
    agent_match = agent_pat.match(user_agent)
    if agent_match:
        return AgentDetails(**agent_match.groupdict())
    return None
```

We've written three parsers for the HTTP request, the time stamp, and the user agent information. The request value in a log is usually a three-word string such as `GET /some/path HTTP/1.1`. The `parse_request()` function extracts these three space-separated values. In the unlikely event that the path has spaces in it, we'll extract the first word and the last word as the method and protocol; all the remaining words are part of the path.

Time parsing is delegated to the `datetime` module. We've provided the proper format in the `parse_time()` function.

Parsing the user agent is challenging. There are many variations; we've chosen a common one for the `parse_agent()` function. If the user agent text matches the given regular expression, we'll use the attributes of the `AgentDetails` class. If the user agent information doesn't match the regular expression, we'll use the `None` value instead. The original text will be available in the `Access` object.

We'll use these three parsers to build AccessDetails instances from the given Access objects. The main body of the access_detail_iter() function looks like this:

```
from typing import Iterable, Iterator
def access_detail_iter(
        access_iter: Iterable[Access]
    ) -> Iterator[AccessDetails]:
    for access in access_iter:
        try:
            meth, url, protocol = parse_request(access.request)
            yield AccessDetails(
                access=access,
                time=parse_time(access.time),
                method=meth,
                url=urllib.parse.urlparse(url),
                protocol=protocol,
                referrer=urllib.parse.urlparse(access.referer),
                agent=parse_agent(access.user_agent)
            )
        except ValueError as e:
            print(e, repr(access))
```

We've used a similar design pattern to the previous access_iter() function. A new object is built from the results of parsing some input object. The new AccessDetails object will wrap the previous Access object. This technique allows us to use immutable objects, yet still contains more refined information.

This function is essentially a mapping from an Access object to an AccessDetails object. Here's an alternative design using the map() higher-level function:

```
from typing import Iterable, Iterator
def access_detail_iter2(
        access_iter: Iterable[Access]
    ) -> Iterator[AccessDetails]:

    def access_detail_builder(access: Access) -> Optional[AccessDetails]:
        try:
            meth, uri, protocol = parse_request(access.request)
            return AccessDetails(
                access=access,
                time=parse_time(access.time),
                method=meth,
                url=urllib.parse.urlparse(uri),
                protocol=protocol,
                referrer=urllib.parse.urlparse(access.referer),
                agent=parse_agent(access.user_agent)
```

```
            )
        except ValueError as e:
            print(e, repr(access))
        return None

    return filter(
        None,
        map(access_detail_builder, access_iter)
    )
```

We've changed the construction of the AccessDetails object to be a function that returns a single value. We can map that function to the iterable input stream of the raw Access objects. We'll see that this fits in nicely with the way the multiprocessing module works.

In an object-oriented programming environment, these additional parsers might be method functions or properties of a class definition. The advantage of an object-oriented design with lazy parsing methods is that items aren't parsed unless they're needed. This particular functional design parses everything, assuming that it's going to be used.

It's possible to create a lazy functional design. It can rely on the three parser functions to extract and parse the various elements from a given Access object as needed. Rather than using the details.time attribute, we'd use the parse_time(access.time) parameter. The syntax is longer, but it ensures that the attribute is only parsed as needed.

Filtering the access details

We'll look at several filters for the AccessDetails objects. The first is a collection of filters that reject a lot of overhead files that are rarely interesting. The second filter will be part of the analysis functions, which we'll look at later.

The path_filter() function is a combination of three functions:

- Exclude empty paths
- Exclude some specific filenames
- Exclude files that have a given extension

An optimized version of the path_filter() function looks like this:

```
def path_filter(
        access_details_iter: Iterable[AccessDetails]
    ) -> Iterable[AccessDetails]:
    name_exclude = {
        'favicon.ico', 'robots.txt', 'index.php', 'humans.txt',
```

```
        'dompdf.php', 'crossdomain.xml',
        '_images', 'search.html', 'genindex.html',
        'searchindex.js', 'modindex.html', 'py-modindex.html',
    }
    ext_exclude = {
        '.png', '.js', '.css',
    }
    for detail in access_details_iter:
        path = detail.url.path.split('/')
        if not any(path):
            continue
        if any(p in name_exclude for p in path):
            continue
        final = path[-1]
        if any(final.endswith(ext) for ext in ext_exclude):
            continue
        yield detail
```

For each individual `AccessDetails` object, we'll apply three filter tests. If the path is essentially empty, if the path includes one of the excluded names, or if the path's final name has an excluded extension, the item is quietly ignored. If the path doesn't match any of these criteria, it's potentially interesting and is part of the results yielded by the `path_filter()` function.

This is an optimization because all of the tests are applied using an imperative style `for` loop body.

An alternative design can define each test as a separate first-class, filter-style function. For example, we might have a function such as the following to handle empty paths:

```
def non_empty_path(detail: AccessDetails) -> bool:
    path = detail.url.path.split('/')
    return any(path)
```

This function simply ensures that the path contains a name. We can use the `filter()` function as follows:

```
filter(non_empty_path, access_details_iter)
```

We can write similar tests for the `non_excluded_names()` and `non_excluded_ext()` functions. The entire sequence of `filter()` functions will look like this:

```
filter(non_excluded_ext,
    filter(non_excluded_names,
        filter(non_empty_path, access_details_iter)))
```

This applies each `filter()` function to the results of the previous `filter()` function. The empty paths are rejected; from this subset, the excluded names and the excluded extensions are rejected. We can also state the preceding example as a series of assignment statements as follows:

```
non_empty = filter(non_empty_path, access_details_iter)
nx_name = filter(non_excluded_names, non_empty)
nx_ext = filter(non_excluded_ext, nx_name)
```

This version has the advantage of being slightly easier to expand when we add new filter criteria.

The use of generator functions (such as the `filter()` function) means that we aren't creating large intermediate objects. Each of the intermediate variables, `ne`, `nx_name`, and `nx_ext`, are proper lazy generator functions; no processing is done until the data is consumed by a client process.

While elegant, this suffers from inefficiency because each function will need to parse the path in the `AccessDetails` object. In order to make this more efficient, we will need to wrap a `path.split('/')` function with the `lru_cache` attribute.

Analyzing the access details

We'll look at two analysis functions that we can use to filter and analyze the individual `AccessDetails` objects. The first function, a `filter()` function, will pass only specific paths. The second function will summarize the occurrences of each distinct path.

We'll define a small `book_in_path()` function and combine this with the built-in `filter()` function to apply the function to the details. Here is the composite `book_filter()` function:

```
from typing import Iterable, Iterator
def book_filter(
        access_details_iter: Iterable[AccessDetails]
    ) -> Iterator[AccessDetails]:
    def book_in_path(detail: AccessDetails) -> bool:
        path = tuple(
            item
            for item in detail.url.path.split('/')
            if item
        )
        return path[0] == 'book' and len(path) > 1
    return filter(book_in_path, access_details_iter)
```

We've defined a rule, through the book_in_path() function, which we'll apply to each AccessDetails object. If the path is not empty and the first-level attribute of the path is book, then we're interested in these objects. All other AccessDetails objects can be quietly rejected.

The reduce_book_total() function is the final reduction that we're interested in:

```
from collections import Counter
def reduce_book_total(
        access_details_iter: Iterable[AccessDetails]
    ) -> Dict[str, int]:
    counts: Dict[str, int] = Counter()
    for detail in access_details_iter:
        counts[detail.url.path] += 1
    return counts
```

This function will produce a Counter() object that shows the frequency of each path in an AccessDetails object. In order to focus on a particular set of paths, we'll use the reduce_total(book_filter(details)) method. This provides a summary of only items that are passed by the given filter.

Because Counter objects can be applied to a wide variety of types, a type hint is required to provide a narrow specification. In this case, the hint is Dict[str, int] to show the **mypy** tool that string representations of paths will be counted.

The complete analysis process

Here is the composite analysis() function that digests a collection of log files:

```
def analysis(filename: str) -> Dict[str, int]:
    """Count book chapters in a given log"""
    details = path_filter(
        access_detail_iter(
            access_iter(
                local_gzip(filename))))
    books = book_filter(details)
    totals = reduce_book_total(books)
    return totalsWe've defined a rule, through the book_in_path() function,
which we'll apply to each
```

The `analysis()` function uses the `local_gzip()` function to work with a single filename or file pattern. It applies a standard set of parsing functions, `path_filter()`, `access_detail_iter()`, and `access_iter()`, to create an iterable sequence of `AccessDetails` objects. It then applies our analytical filter and reduction to that sequence of `AccessDetails` objects. The result is a `Counter` object that shows the frequency of access for certain paths.

A specific collection of saved `.gzip` format log files totals about 51 MB. Processing the files serially with this function takes over 140 seconds. Can we do better using concurrent processing?

Using a multiprocessing pool for concurrent processing

One elegant way to make use of the `multiprocessing` module is to create a processing `Pool` object and assign work to the various processes in that pool. We will use the OS to interleave execution among the various processes. If each of the processes has a mixture of I/O and computation, we should be able to ensure that our processor is very busy. When processes are waiting for the I/O to complete, other processes can do their computations. When an I/O finishes, a process will be ready to run and can compete with others for processing time.

The recipe for mapping work to a separate process looks like this:

```
import multiprocessing
    with multiprocessing.Pool(4) as workers:
        workers.map(analysis, glob.glob(pattern))
```

We've created a `Pool` object with four separate processes and assigned this `Pool` object to the `workers` variable. We've then mapped a function, `analysis`, to an iterable queue of work to be done, using the pool of processes. Each process in the `workers` pool will be assigned items from the iterable queue. In this case, the queue is the result of the `glob.glob(pattern)` attribute, which is a sequence of file names.

As the `analysis()` function returns a result, the parent process that created the `Pool` object can collect those results. This allows us to create several concurrently built `Counter` objects and to merge them into a single, composite result.

If we start *p* processes in the pool, our overall application will include *p+1* processes. There will be one parent process and *p* children. This often works out well because the parent process will have little to do after the subprocess pools are started. Generally, the workers will be assigned to separate CPUs (or cores) and the parent will share a CPU with one of the children in the `Pool` object.

> The ordinary Linux parent/child process rules apply to the subprocesses created by this module. If the parent crashes without properly collecting the final status from the child processes, then *zombie* processes can be left running. For this reason, a process `Pool` object is a context manager. When we use a pool through the `with` statement, at the end of the context, the children are properly terminated.

By default, a `Pool` object will have a number of workers based on the value of the `multiprocessing.cpu_count()` function. This number is often optimal, and simply using the `with multiprocessing.Pool() as workers:` attribute might be sufficient.

In some cases, it can help to have more workers than CPUs. This might be true when each worker has I/O-intensive processing. Having many worker processes waiting for I/O to complete can improve the elapsed running time of an application.

If a given `Pool` object has *p* workers, this mapping can cut the processing time to almost $\frac{1}{p}$ of the time required to process all of the logs serially. Pragmatically, there is some overhead involved with communication between the parent and child processes in the `Pool` object. These overheads will limit the effectiveness of subdividing the work into very small concurrent pieces.

The multiprocessing `Pool` object has four map-like methods to allocate work to a pool: `map()`, `imap()`, `imap_unordered()`, and `starmap()`. Each of these is a variation of the common theme of assigning a function to a pool of processes and mapping data items to that function. They differ in the details of allocating work and collecting results.

The `map(function, iterable)` method allocates items from the iterable to each worker in the pool. The finished results are collected in the order they were allocated to the `Pool` object so that order is preserved.

The `imap(function, iterable)` method is described as `lazier` than `map()`. By default, it sends each individual item from the iterable to the next available worker. This might involve more communication overhead. For this reason, a chunk size larger than one is suggested.

The imap_unordered(function, iterable) method is similar to the imap() method, but the order of the results is not preserved. Allowing the mapping to be processed out of order means that, as each process finishes, the results are collected. Otherwise, the results must be collected in order.

The starmap(function, iterable) method is similar to the itertools.starmap() function. Each item in the iterable must be a tuple; the tuple is passed to the function using the * modifier so that each value of the tuple becomes a positional argument value. In effect, it's performing function(*iterable[0]), function(*iterable[1]), and so on.

Here is one of the variations of the preceding mapping theme:

```
import multiprocessing
pattern = "*.gz"
combined = Counter()
with multiprocessing.Pool() as workers:
    result_iter = workers.imap_unordered(
        analysis, glob.glob(pattern))
    for result in result_iter:
        combined.update(result)
```

We've created a Counter() function that we'll use to consolidate the results from each worker in the pool. We created a pool of subprocesses based on the number of available CPUs and used the Pool object as a context manager. We then mapped our analysis() function to each file in our file-matching pattern. The resulting Counter objects from the analysis() function are combined into a single resulting counter.

This version took about 68 seconds to analyze a batch of log files. The time to analyze the logs was cut dramatically using several concurrent processes. The single-process baseline time was 150 seconds. Other experiments need to be run with larger pool sizes to determine how many workers are required to make the system as busy as possible.

We've created a two-tiered map-reduce process with the multiprocessing module's Pool.map() function. The first tier was the analysis() function, which performed a map-reduce on a single log file. We then consolidated these reductions in a higher-level reduce operation.

Using apply() to make a single request

In addition to the `map()` function's variants, a pool also has an `apply(function, *args, **kw)` method that we can use to pass one value to the worker pool. We can see that the `map()` method is really just a `for` loop wrapped around the `apply()` method. We can, for example, use the following command:

```
list(
    workers.apply(analysis, f)
    for f in glob.glob(pattern)
)
```

It's not clear, for our purposes, that this is a significant improvement. Almost everything we need to do can be expressed as a `map()` function.

Using the map_async(), starmap_async(), and apply_async() functions

The role of the `map()`, `starmap()`, and `apply()` functions is to allocate work to a subprocess in the `Pool` object and then collect the response from the subprocess when that response is ready. This can cause the child to wait for the parent to gather the results. The `_async()` function's variations do not wait for the child to finish. These functions return an object that can be queried to get the individual results from the child processes.

The following is a variation using the `map_async()` method:

```
import multiprocessing
pattern = "*.gz"
combined = Counter()
with multiprocessing.Pool() as workers:
    results = workers.map_async(
        analysis, glob.glob(pattern))
    data = results.get()
    for c in data:
        combined.update(c)
```

We've created a `Counter()` function that we'll use to consolidate the results from each worker in the pool. We created a pool of subprocesses based on the number of available CPUs and used this `Pool` object as a context manager. We then mapped our `analysis()` function to each file in our file-matching pattern. The response from the `map_async()` function is a `MapResult` object; we can query this for the results and overall status of the pool of workers. In this case, we used the `get()` method to get the sequence of the `Counter` objects.

The resulting `Counter` objects from the `analysis()` function are combined into a single resulting `Counter` object. This aggregate gives us an overall summary of a number of log files. This processing is not any faster than the previous example. The use of the `map_async()` function allows the parent process to do additional work while waiting for the children to finish.

More complex multiprocessing architectures

The `multiprocessing` package supports a wide variety of architectures. We can easily create multiprocessing structures that span multiple servers and provide formal authentication techniques to create a necessary level of security. We can pass objects from process to process using queues and pipes. We can share memory between processes. We can also share lower-level locks between processes as a way to synchronize access to shared resources such as files.

Most of these architectures involve explicitly managing states among several working processes. Using locks and shared memory, in particular, is imperative in nature and doesn't fit in well with a functional programming approach.

We can, with some care, treat queues and pipes in a functional manner. Our objective is to decompose a design into producer and consumer functions. A producer can create objects and insert them into a queue. A consumer will take objects out of a queue and process them, perhaps putting intermediate results into another queue. This creates a network of concurrent processors and the workload is distributed among these various processes. Using the `pycsp` package can simplify the queue-based exchange of messages among processes. For more information, visit `https://pypi.python.org/pypi/pycsp`.

This design technique has some advantages when designing a complex application server. The various subprocesses can exist for the entire life of the server, handling individual requests concurrently.

Using the concurrent.futures module

In addition to the `multiprocessing` package, we can also make use of the `concurrent.futures` module. This also provides a way to map data to a concurrent pool of threads or processes. The module API is relatively simple and similar in many ways to the `multiprocessing.Pool()` function's interface.

Here is an example to show just how similar they are:

```
from concurrent.futures import ProcessPoolExecutor
pool_size = 4
pattern = "*.gz"
combined = Counter()
with ProcessPoolExecutor(max_workers=pool_size) as workers:
    for result in workers.map(analysis, glob.glob(pattern)):
        combined.update(result)
```

The most significant change between the preceding example and previous examples is that we're using an instance of the `concurrent.futures.ProcessPoolExecutor` object instead of the `multiprocessing.Pool` method. The essential design pattern is to map the `analysis()` function to the list of filenames using the pool of available workers. The resulting `Counter` objects are consolidated to create a final result.

The performance of the `concurrent.futures` module is nearly identical to the `multiprocessing` module.

Using concurrent.futures thread pools

The `concurrent.futures` module offers a second kind of executor that we can use in our applications. Instead of creating a `concurrent.futures.ProcessPoolExecutor` object, we can use the `ThreadPoolExecutor` object. This will create a pool of threads within a single process.

The syntax for thread pools is almost identical to using a `ProcessPoolExecutor` object. The performance, however, is remarkably different. In this example of log file analysis, the work is dominated by I/O. Because all of the threads in a process share the same OS scheduling constraints, the overall performance of multithreaded log file analysis is about the same as processing the log files serially.

Using sample log files and a small four-core laptop running macOS X, these are the kinds of results that indicate the difference between threads that share I/O resources and processes:

- Using the `concurrent.futures` thread pool, the elapsed time was 168 seconds
- Using a process pool, the elapsed time was 68 seconds

In both cases, the `Pool` object's size was 4. The single-process and single-thread baseline time was 150 seconds; adding threads made processing run more slowly. This result is typical of programs doing a great deal of input and output. Multithreading may be more appropriate for user interfaces where threads are idle for long periods of time, or where waiting for the person to move the mouse or touch the screen.

Using the threading and queue modules

The Python `threading` package involves a number of constructs helpful for building imperative applications. This module is not focused on writing functional applications. We can make use of thread-safe queues in the `queue` module to pass objects from thread to thread.

The `threading` module doesn't have a simple way of distribuing work to various threads. The API isn't ideally suited to functional programming.

As with the more primitive features of the `multiprocessing` module, we can try to conceal the stateful and imperative nature of locks and queues. It seems easier, however, to make use of the `ThreadPoolExecutor` method in the `concurrent.futures` module. The `ProcessPoolExecutor.map()` method provides us with a very pleasant interface to concurrently process the elements of a collection.

The use of the `map()` function primitive to allocate work seems to fit nicely with our functional programming expectations. For this reason, it's best to focus on the `concurrent.futures` module as the most accessible way to write concurrent functional programs.

Designing concurrent processing

From a functional programming perspective, we've seen three ways to use the `map()` function concept applied to data items concurrently. We can use any one of the following:

- `multiprocessing.Pool`
- `concurrent.futures.ProcessPoolExecutor`
- `concurrent.futures.ThreadPoolExecutor`

These are almost identical in the way we interact with them; all three have a `map()` method that applies a function to items of an iterable collection. This fits in elegantly with other functional programming techniques. The performance is different because of the nature of concurrent threads versus concurrent processes.

As we stepped through the design, our log analysis application decomposed into two overall areas:

- The lower-level parsing: This is generic parsing that will be used by almost any log analysis application
- The higher-level analysis application: This is more specific filtering and reduction focused on our application needs

The lower-level parsing can be decomposed into four stages:

- Reading all the lines from multiple source log files. This was the `local_gzip()` mapping from file name to a sequence of lines.
- Creating simple namedtuples from the lines of log entries in a collection of files. This was the `access_iter()` mapping from text lines to `Access` objects.
- Parsing the details of more complex fields such as dates and URLs. This was the `access_detail_iter()` mapping from `Access` objects to `AccessDetails` objects.
- Rejecting uninteresting paths from the logs. We can also think of this as passing only the interesting paths. This was more of a filter than a map operation. This was a collection of filters bundled into the `path_filter()` function.

We defined an overall `analysis()` function that parsed and analyzed a given log file. It applied the higher-level filter and reduction to the results of the lower-level parsing. It can also work with a wildcard collection of files.

Given the number of mappings involved, we can see several ways to decompose this problem into work that can be mapped to into a pool of threads or processes. Here are some of the mappings we can consider as design alternatives:

- Map the `analysis()` function to individual files. We use this as a consistent example throughout this chapter.
- Refactor the `local_gzip()` function out of the overall `analysis()` function. We can now map the revised `analysis()` function to the results of the `local_gzip()` function.
- Refactor the `access_iter(local_gzip(pattern))` function out of the overall `analysis()` function. We can map this revised `analysis()` function against the iterable sequence of the `Access` objects.
- Refactor the `access_detail_iter(access-iter(local_gzip(pattern)))` function into a separate iterable. We will then map the `path_filter()` function and the higher-level filter and reduction against the iterable sequence of the `AccessDetail` objects.
- We can also refactor the lower-level parsing into a function that is separate from the higher-level analysis. We can map the analysis filter and reduction against the output from the lower-level parsing.

All of these are relatively simple methods to restructure the example application. The benefit of using functional programming techniques is that each part of the overall process can be defined as a mapping. This makes it practical to consider different architectures to locate an optimal design.

In this case, however, we need to distribute the I/O processing to as many CPUs or cores as we have available. Most of these potential refactorings will perform all of the I/O in the parent process; these will only distribute the computations to multiple concurrent processes with little resulting benefit. Then, we want to focus on the mappings, as these distribute the I/O to as many cores as possible.

It's often important to minimize the amount of data being passed from process to process. In this example, we provided just short filename strings to each worker process. The resulting `Counter` object was considerably smaller than the 10 MB of compressed detail data in each log file. We can further reduce the size of each `Counter` object by eliminating items that occur only once, or we can limit our application to only the 20 most popular items.

The fact that we can reorganize the design of this application freely doesn't mean we should reorganize the design. We can run a few benchmarking experiments to confirm our suspicions that log file parsing is dominated by the time required to read the files.

Summary

In this chapter, we've looked at two ways to support the concurrent processing of multiple pieces of data:

- The `multiprocessing` module: Specifically, the `Pool` class and the various kinds of mappings available to a pool of workers.
- The `concurrent.futures` module: Specifically, the `ProcessPoolExecutor` and `ThreadPoolExecutor` classes. These classes also support a mapping that will distribute work among workers that are threads or processes.

We've also noted some alternatives that don't seem to fit in well with functional programming. There are numerous other features of the `multiprocessing` module, but they're not a good fit with functional design. Similarly, the `threading` and `queue` modules can be used to build multithreaded applications, but the features aren't a good fit with functional programs.

In the next chapter, we'll look at the `operator` module. This can be used to simplify some kinds of algorithm. We can use a built-in operator function instead of defining a Lambda form. We'll also look at some techniques to design flexible decision making and to allow expressions to be evaluated in a non-strict order.

13
Conditional Expressions and the Operator Module

Functional programming emphasizes lazy or non-strict ordering of operations. The idea is to allow the compiler or runtime to do as little work as possible to compute the answer. Python tends to impose strict ordering on evaluations, which could be inefficient.

The Python `if`, `elif`, and `else` statements enforce a strict ordering on the evaluation of the conditions. In this chapter, we'll look at ways we can, to an extent, free ourselves from the strict ordering, and develop a limited kind of non-strict conditional statement. It's not clear whether this is helpful, but it will show some alternative ways to express an algorithm in a somewhat more functional style.

In the previous chapters, we looked at a number of higher-order functions. In some cases, we used these higher-order functions to apply fairly sophisticated functions to collections of data. In other cases, we applied simple functions to collections of data.

Indeed, in many cases, we wrote tiny lambda objects to apply a single Python operator to a function. For example, we can use the following code to define a `prod()` function:

```
from typing import Iterable
from functools import reduce
def prod(data: Iterable[int]) -> int:
    return reduce(lambda x, y: x*y, data, 1)
```

The use of the `lambda x,y: x*y` parameter seems a bit wordy for simple multiplication. After all, we just want to use the multiplication operator (`*`). Can we simplify the syntax? The answer is yes; the `operator` module provides us with definitions of the built-in operators. In this case, we can use `operator.mul` instead of the lambda object.

We'll look at several topics in this chapter:

- The first part will look at ways we can implement non-strict evaluation. This is a tool that's interesting because it can lead to performance optimizations.
- We'll also look at the `operator` module and how this leads to some simplification and potential clarification to create higher-order functions.
- We'll look at star-mapping, where a `f(*args)` is used to provide multiple arguments to a mapping.
- We'll look at some more advanced `partial()` and `reduce()` techniques as well.

Evaluating conditional expressions

Python imposes strict ordering on expressions; the notable exceptions are the short-circuit operators, `and` and `or`. It imposes strict ordering on statement evaluation. This makes it challenging to find optimizations because they would break the strict evaluation order.

Evaluating condition expressions is one way in which we can experiment with non-strict ordering of statements. The `if`, `elif`, and `else` Python statements are evaluated in a strict order from first to last. Ideally, an optimizing language may relax this rule so that a compiler can find a faster order for evaluating the conditional expressions. This allows us to write the expressions in an order that makes sense to a reader, and lets the compiler find a faster evaluation order.

Lacking an optimizing compiler, the concept of non-strict ordering is a bit of a stretch for Python. Nonetheless, we do have alternative ways to express conditions that involve the evaluation of functions instead of the execution of imperative statements. This can allow some re-arrangement at runtime.

Python has a conditional `if` and `else` expression. The `if-else` operator is a short-circuit operator. It leads to a tiny optimization because only one of the two outside conditions is evaluated based on the truthfulness of the inner condition. When we write `x if c else y`, the `x` expression is only evaluated if `c` is `True`. Additionally, the `y` expression is only evaluated if `c` is `False`. This is a minor optimization, but the order of operations is still strictly enforced.

This expression is useful for simple conditions. When we have multiple conditions, however, it can get awkwardly complex; we'd have to carefully nest the subexpressions. We might wind up with an expression, as follows, which is rather difficult to comprehend:

```
(x if n==1 else (y if n==2 else z))
```

The preceding expression will only evaluate one of `x`, `y`, or `z`, depending on the value of `n`.

When looking at the `if` statement, we have some data structures that can mimic the effect of the `if` statement. One technique is to use dictionary keys and lambda objects to create a mapping set of conditions and values. Here's a way to express the Factorial function as expressions:

```
def fact(n: int) -> int:
    f = {
        n == 0: lambda n: 1,
        n == 1: lambda n: 1,
        n == 2: lambda n: 2,
        n > 2: lambda n: fact(n-1)*n
    }[True]
    return f(n)
```

This rewrites the conventional `if`, `elif`, `elif`, and `else` sequence of statements into a single expression. We've decomposed it into two steps to make what's happening slightly clearer.

In the first step, we'll evaluate the various conditions. One of the given conditions will evaluate to `True`, the others should all evaluate to `False`. The resulting dictionary will have two items in it—a `True` key with a lambda object and a `False` key with a lambda object. We'll select the `True` item and assign it to the `f` variable.

We used lambdas as the values in this mapping so that the value expressions aren't evaluated when the dictionary is built. We want the dictionary to pick a lambda. This lambda's value is the result of the function overall. The `return` statement evaluates the one lambda associated with the `True` condition, `f`, applied to the input argument, `n`.

Exploiting non-strict dictionary rules

Prior to Python 3.6, a dictionary's keys had no defined order. If we try to create a dictionary with multiple items that share a common key value, there will only be one item in the resulting `dict` object. There was no definition for which of the duplicated key values will be preserved, and if we've designed the algorithm properly, it shouldn't matter.

The following result is typical. The last value has replaced any prior values. Prior to Python 3.6, there was no guarantee this would happen:

```
>>> {'a': 1, 'a': 2}
{'a': 2}
```

Here's a situation where we explicitly don't care which of the duplicated keys is preserved. If we look at a degenerate case of the `max()` function, it simply picks the largest of two values:

```
def non_strict_max(a, b):
    f = {a >= b: lambda: a,
         b >= a: lambda: b}[True]
    return f()
```

In the case where `a == b`, both items in the dictionary will have a key of the `True` condition. Only one of the two will actually be preserved. Since the answer is the same, it doesn't matter which is kept, and which is treated as a duplicate and overwritten.

Note that a formal type hint for this function is surprisingly complex. The items which are being compared must be *rankable*—they must implement the ordering operators. Here's how we can define a type that captures the idea of rankable objects:

```
from abc import ABCMeta, abstractmethod
from typing import TypeVar, Any

class Rankable(metaclass=ABCMeta):
    @abstractmethod
    def __lt__(self, other: Any) -> bool: ...
    @abstractmethod
    def __gt__(self, other: Any) -> bool: ...
    @abstractmethod
    def __le__(self, other: Any) -> bool: ...
    @abstractmethod
    def __ge__(self, other: Any) -> bool: ...

RT = TypeVar('RT', bound=Rankable)
```

The `Rankable` class definition is an abstract class, and relies on the `abc` module for some useful definitions to formalize the abstract nature of the class definition. The `@abstractmethod` decorators are used to identify method functions which must be defined by any useful, concrete subclass.

A typevariable, RT, can then be bound to the parameters and result type of the non_strict_max() function. Here's the definition with the type hints included (the body has been omitted, as it's shown earlier):

```
def non_strict_max(a: RT, b: RT) -> RT:
```

This clarifies that the two parameters, a, and b, are expected to be of some rankable type which will be bound to RT. The result will be of the same rankable type. This clarifies the usual semantics of max() where the arguments are of a consistent type and the result is of the same type.

Filtering true conditional expressions

We have a number of ways of determining which expression is True. In the previous example, we loaded the keys into a dictionary. Because of the way the dictionary is loaded, only one value will be preserved with a key of True.

Here's another variation to this theme, written using the filter() function:

```
from operator import itemgetter
def semifact(n: int) -> int:
    alternatives = [
        (n == 0, lambda n: 1),
        (n == 1, lambda n: 1),
        (n == 2, lambda n: 2),
        (n > 2, lambda n: semifact(n-2)*n)
    ]
    _, f = next(filter(itemgetter(0), alternatives))
    return f(n)
```

We defined the alternatives as a sequence of condition and function pairs. Each item in the as a condition based on the input and a lambda item that will produce the output. We could also include a type hint in the variable assignment to look like this:

```
alternatives: List[Tuple[bool, Callable[[int], int]]] = [
   etc,
]
```

The list is literally the same collection of four two-tuples. This definition clarifies that the list of tuples are a Boolean result and a callable function.

When we apply the `filter()` function using the `itemgetter(0)` parameter, we'll select those pairs with a `True` value in item zero of the tuple. Of those which are `True`, we'll use `next()` to extract the first item from the iterable created by the `filter()` function. The selected condition value is assigned to the _ variable; the selected function is assigned to the `f` variable. We can ignore the condition value (it will be `True`), and we can evaluate the `f()` function that was returned.

As with the previous example, we used lambdas to defer evaluation of the functions until after the conditions have been evaluated.

This `semifact()` function is also called **double factorial**. The definition of semifactorial is similar to the definition of factorial. The important difference is that it is the product of alternate numbers instead of all numbers. For examples, take a look at the following formulas:

- $5!! = 5 \times 3 \times 1$
- $7!! = 7 \times 5 \times 3 \times 1$

Finding a matching pattern

This technique of creating a collection of multiple conditions can also be applied when using regular expressions. Recall that a pattern's `match()` or `search()` method either returns a match object or `None`. We can leverage this with programs, as shown in the following code:

```
import re
p1 = re.compile(r"(some) pattern")
p2 = re.compile(r"a (different) pattern")

from typing import Optional, Match
def matcher(text: str) -> Optional[Match[str]]:
    patterns = [p1, p2]
    matching = (p.search(text) for p in patterns)
    try:
        good = next(filter(None, matching))
        return good
    except StopIteration:
        pass
```

We've defined two patterns that we'd like to apply against a given block of text. Each pattern has a sub-pattern marked with `()`, which will be a capture group.

The `matcher()` function will build a sequence of alternative patterns. In this case, it's a simple literal pair of patterns. A generator expression is used to apply each pattern's `search()` method against the supplied text. Because generator expressions are lazy, this doesn't immediately perform a long series of pattern matches; instead, the results must be consumed.

The `filter()` function with `None` as the first argument will remove all `None` values from the sequence of items. The value of `filter(None, S)` is the same as `filter(lambda item: item is not None, S)`.

The `next()` function will take the first non-none value from the iterable results of the `filter()` function. If there are no results from the `filter()` function it means no pattern matched. In this case, we transformed the exception into a `None` result. A sensible alternative may be to raise a custom exception because no pattern matched the given text.

As with the previous examples, this shows how a number of Boolean conditions can be evaluated and a single true value selected. Because the input is a sequence of patterns, the order of evaluation of the various functions is well-defined, and performed in a strict order. While we can't escape from Python's strict order of evaluation, we can limit the cost of evaluation by exiting from the function as soon as a match was found.

Using the operator module instead of lambdas

When using the `max()`, `min()`, and `sorted()` functions, we have an optional `key=` parameter. The function provided as an argument value modifies the behavior of the higher-order function. In many cases, we used simple lambda forms to pick items from a tuple. Here are two examples we heavily relied on:

```
from typing import Callable, Sequence, TypeVar
T_ = TypeVar("T_")
fst: Callable[[Sequence[T_]], T_] = lambda x: x[0]
snd: Callable[[Sequence[T_]], T_] = lambda x: x[1]
```

These match built-in functions in some other functional programming languages that are used to pick the first or second item from a tuple. This includes type hints to assure that there's no other transformation going on—the type of items in the sequence is bound to the type variable, `T_`, which reflects the type of the result of the function.

We don't really need to write these functions. There's a version available in the `operator` module named `itemgetter()`. These are higher-order functions. The expression `itemgetter(0)` creates a function. We can then apply the function to a collection to select an object, like this.

```
>>> from operator import itemgetter
>>> itemgetter(0)([1, 2, 3])
1
```

Let's use this on a more complex list of tuples data structure. Here's some sample data we can work with:

```
year_cheese = [
    (2000, 29.87), (2001, 30.12), (2002, 30.6), (2003, 30.66),
    (2004, 31.33), (2005, 32.62), (2006, 32.73), (2007, 33.5),
    (2008, 32.84), (2009, 33.02), (2010, 32.92)
]
```

This is the annual cheese consumption. We used this example in `Chapter 2`, *Introducing Essential Functional Concepts*, and `Chapter 9`, *More Itertools Techniques*.

We can locate the data point with minimal cheese using the following commands:

```
>>> min(year_cheese, key=snd)
(2000, 29.87)
```

The `operator` module gives us an alternative to pick particular elements from a tuple. This saves us from using a lambda variable to pick the second item.

Instead of defining our own `fst()` and `snd()` functions, we can use the `itemgetter(0)` and the `itemgetter(1)` parameters, as shown in the following command:

```
>>> from operator import itemgetter
>>> max(year_cheese, key=itemgetter(1))
(2007, 33.5)
```

The `itemgetter()` function relies on the special method, `__getitem__()`, to pick items out of a tuple (or list) based on their index position.

Getting named attributes when using higher-order functions

Let's look at a slightly different collection of data. Let's say we were working with NamedTuple subclasses instead of anonymous tuples. First, we'll define a class that has type hints for both items within the tuple:

```
from typing import NamedTuple
class YearCheese(NamedTuple):
    year: int
    cheese: float
```

Then, we can convert our base year_cheese data into properly named tuples. The conversion is shown, as follows:

```
>>> year_cheese_2 = list(YearCheese(*yc) for yc in year_cheese)

>>> year_cheese_2
[YearCheese(year=2000, cheese=29.87),
 YearCheese(year=2001, cheese=30.12),
 YearCheese(year=2002, cheese=30.6),
 YearCheese(year=2003, cheese=30.66),
 YearCheese(year=2004, cheese=31.33),
 YearCheese(year=2005, cheese=32.62),
 YearCheese(year=2006, cheese=32.73),
 YearCheese(year=2007, cheese=33.5),
 YearCheese(year=2008, cheese=32.84),
 YearCheese(year=2009, cheese=33.02),
 YearCheese(year=2010, cheese=32.92)]
```

We have two ways to locate the range of cheese consumption. We can use the attrgetter() function, or we can use a lambda form, as follows:

```
>>> from operator import attrgetter
>>> min(year_cheese_2, key=attrgetter('cheese'))
YearCheese(year=2000, cheese=29.87)
>>> max(year_cheese_2, key=lambda x: x.cheese)
YearCheese(year=2007, cheese=33.5)
```

What's important here is that, with a lambda object, the attribute name is expressed as a token in the code. With the attrgetter() function, the attribute name is a character string. Using a character string the attribute name could be a parameter which can be changed when the script is run, allowing us some additional flexibility.

Starmapping with operators

The itertools.starmap() function is a variation on the map() higher-order function. The map() function applies a function against each item from a sequence. The starmap(f, S) function presumes each item, i, from the sequence, S, is a tuple, and uses f(*i). The number of items in each tuple must match the number of parameters in the given function.

Here's an example:

```
>>> d = starmap(pow, zip_longest([], range(4), fillvalue=60))
```

The itertools.zip_longest() function will create a sequence of pairs, such as the following:

```
[(60, 0), (60, 1), (60, 2), (60, 3)]
```

It does this because we provided two sequences: the [] brackets and the range(4) parameter. The fillvalue parameter will be used when the shorter sequence runs out of data.

When we use the starmap() function, each pair becomes the argument to the given function. In this case, we provided the operator.pow() function, which is the ** operator. This expression calculates values for [60**0, 60**1, 60**2, 60**3]. The value of the d variable is [1, 60, 3600, 216000].

The starmap() function expects a sequence of tuples. We have a tidy equivalence between the map(f, x, y) and starmap(f, zip(x, y)) functions.

Here's a continuation of the preceding example of the itertools.starmap() function:

```
>>> p = (3, 8, 29, 44)
>>> pi = sum(starmap(truediv, zip(p, d)))
```

We've zipped together two sequences of four values. The value of the d variable was computed above using starmap(). The p variable refers to a simple list of literal items. We zipped these to make pairs of items. We used the starmap() function with the operator.truediv() function, which is the / operator.
This will compute a sequence of fractions that we sum. The sum is an approximation of $\pi \approx \frac{3}{60^0} + \frac{8}{60^1} + \frac{29}{60^2} + \frac{44}{60^3}$.

Here's a slightly simpler version that uses the `map(f, x, y)` function instead of the `starmap(f, zip(x,y))` function:

```
>>> pi = sum(map(truediv, p, d))
>>> pi
3.1415925925925925
```

In this example, we effectively converted a base `60` fractional value to base `10`. The sequence of values in the `d` variable are the appropriate denominators. A technique similar to the one explained earlier in this section can be used to convert other bases.

Some approximations involve potentially infinite sums (or products). These can be evaluated using similar techniques explained previously in this section. We can leverage the `count()` function in the `itertools` module to generate an arbitrary number of terms in an approximation. We can then use the `takewhile()` function to only accumulate values that contribute a useful level of precision to the answer. Looked at another way, `takewhile()` yields a stream of significant values, and stops processing when an insignificant value is found.

Here's an example of computing a sum from a potentially infinite sequence:

```
>>> from itertools import count, takewhile
>>> num = map(fact, count())
>>> den = map(semifact, (2*n+1 for n in count()))
>>> terms = takewhile(
... lambda t: t > 1E-10, map(truediv, num, den))
>>> 2*sum(terms)
3.1415926533011587
```

The `num` variable is a potentially infinite sequence of numerators, based on a factorial function, defined in a previous example. The `count()` function returns ascending values, starting from zero and continuing indefinitely. The `den` variable is a potentially infinite sequence of denominators, based on the semifactorial (sometimes called the double factorial) function. This function is also defined in a previous example. It also uses `count()` to create a potentially infinite series of values.

To create terms, we used the `map()` function to apply the `operators.truediv()` function, the / operator, to each pair of values. We wrapped this in a `takewhile()` function so that we only take terms from the `map()` output while the value is greater than some relatively small value; in this case, 10^{-10}.

This is a series expansion based on this definition:

$$4\arctan(1) = \pi = 2\sum_{n=0}^{\infty} \frac{n!}{(2n+1)!!}$$

An interesting variation to the series expansion theme is to replace the `operator.truediv()` function with the `fractions.Fraction()` function. This will create exact rational values that don't suffer from the limitations of floating-point approximations.

All of the built-in Python operators are available in the `operators` module. This includes all of the bit-fiddling operators as well as the comparison operators. In some cases, a generator expression may be more succinct or expressive than a rather complex-looking `starmap()` function with a function that represents an operator.

The `operator` module functions a shorthand for a lambda. We can use the `operator.add` method instead of the `add=lambda a, b: a+b` method. If we have expressions more complex than a single operator, then the lambda object is the only way to write them.

Reducing with operator module functions

We'll look at one more way that we can use the operator definitions. We can use them with the built-in `functools.reduce()` function. The `sum()` function, for example, can be defined as follows:

```
sum = functools.partial(functools.reduce, operator.add)
```

This creates a partially evaluated version of the `reduce()` function with the first argument supplied. In this case, it's the + operator, implemented via the `operator.add()` function.

If we have a requirement for a similar function that computes a product, we can define it like this:

```
prod = functools.partial(functools.reduce, operator.mul)
```

This follows the pattern shown in the preceding example. We have a partially evaluated `reduce()` function with the first argument of the `*` operator, as implemented by the `operator.mul()` function.

It's not clear whether we can do similar things with too many of the other operators. We might be able to find a use for the `operator.concat()` function, as well as the `operator.and()` and `operator.or()` functions.

The `and()` and `or()` functions are the bit-wise `&` and `/` operators. If we want the proper Boolean operators, we have to use the `all()` and `any()` functions instead of the `reduce()` function.

Once we have a `prod()` function, this means that the factorial can be defined as follows:

```
fact = lambda n: 1 if n < 2 else n*prod(range(1, n))
```

This has the advantage of being succinct: it provides a single-line definition of factorial. It also has the advantage of not relying on recursion and avoids any problem with stack limitations.

It's not clear that this has any dramatic advantages over the many alternatives we have in Python. The concept of building a complex function from primitive pieces such as the `partial()` and `reduce()` functions and the `operator` module is very elegant. In most cases, though, the simple functions in the `operator` module aren't very helpful; we'll almost always want to use more complex lambdas.

For Boolean reductions, we have to use the built-in `any()` and `all()` functions. These implement a kind of short-circuit `reduce()` operation. They aren't higher-order functions, and they must be used with lambdas or defined functions.

Summary

In this chapter, we looked at alternatives to the `if`, `elif`, and `else` statement sequence. Ideally, using a conditional expression allows some optimization to be done. Pragmatically, Python doesn't optimize, so there's little tangible benefit to the more exotic ways to handle conditions.

We also looked at how we can use the `operator` module with higher-order functions such as `max()`, `min()`, `sorted()`, and `reduce()`. Using operators can save us from having to create a number of small lambdas.

In the next chapter, we'll look at the `PyMonad` library to express a functional programming concept directly in Python. We don't require monads generally because Python is an imperative programming language under the hood.

Some algorithms might be expressed more clearly with monads than with stateful variable assignments. We'll look at an example where monads lead to a succinct expression of a rather complex set of rules. Most importantly, the `operator` module shows off many functional programming techniques.

14
The PyMonad Library

A monad allows us to impose an order on an expression evaluation in an otherwise lenient language. We can use a monad to insist that an expression such as $a + b + c$ is evaluated in left-to-right order. This can interfere with the compiler's ability to optimize expression evaluation. This is necessary, for example, when we want files to have their content read or written in a specific order: a monad assures that the `read()` and `write()` functions are evaluated in a particular order.

Languages that are lenient and have optimizing compilers benefit from monads imposing order on the evaluation of expressions. Python, for the most part, is strict and does not optimize. There are no practical requirements for monads.

However, the PyMonad package contains more than just monads. There are a number of functional programming features that have a distinctive implementation. In some cases, the PyMonad module can lead to programs which are more succinct and expressive than those written using only the standard library modules.

In this chapter, we'll look at the following:

- Downloading and installing PyMonad
- The idea of currying and how this applies to functional composition
- The PyMonad star operator for creating composite functions
- Functors and techniques for currying data items with more generalized functions
- The `bind()` operation, using >>, to create ordered monads
- We'll also explain how to build Monte Carlo simulations using PyMonad techniques

Downloading and installing

The PyMonad package is available on the **Python Package Index** (**PyPi**). In order to add PyMonad to your environment, you'll need to use `pip`.

Visit `https://pypi.python.org/pypi/PyMonad` for more information.

For Mac OS and Linux developers, the `pip install pymonad` command may require the `sudo` command as a prefix. If you've installed a personal version of Python, you won't need to use `sudo`. If you've installed Python system-wide, then you'll need `sudo`. When running a command such as `sudo pip install pymonad`, you'll be prompted for your password to ensure that you have the administrative permissions necessary to do the installation. For Windows developers, the `sudo` command is not relevant, but you do need to have administrative rights.

Once the PyMonad package is installed, you can confirm it using the following commands:

```
>>> import pymonad
>>> help(pymonad)
```

This will display the `docstring` module and confirm that things really are properly installed.

The overall project name, PyMonad, uses mixed case. The installed Python package name that we import, PyMonad, is all lower case. Currently, there are no type hints in the PyMonad package. The very generic nature of the features would require extensive use of `TypeVar` hints to describe the signatures of the various functions. Additionally, there are a few names in PyMonad that conflict with names in the built-in `typing` module; this makes the syntax for type hints potentially more complex because the package name is required to disambiguate the overloaded names. We'll omit type hints for the examples in this chapter.

Functional composition and currying

Some functional languages work by transforming a multiargument function syntax into a collection of single argument functions. This process is called **currying**: it's named after logician Haskell Curry, who developed the theory from earlier concepts.

Currying is a technique for transforming a multiargument function into higher-order single argument functions. In a simple case, consider a function $f(x, y) \rightarrow z$; given two arguments *x* and *y*; this will return some resulting value, *z*. We can curry $f(x, y)$ into into two functions: $f_{c1}(x) \rightarrow f_{c2}(y)$ and $f_{c2}(y) \rightarrow z$. Given the first argument value, *x*, evaluating the function $f_{c1}(x)$ returns a new one-argument function, $f_{c2}(y)$. This second function can be given the second argument value, *y*, and it returns the desired result, *z*.

We can evaluate a curried function in Python as follows: `f_c(2)(3)`. We apply the curried function to the first argument value of `2`, creating a new function. Then, we apply that new function to the second argument value of `3`.

This applies to functions of any complexity. If we start with a function $g(a, b, c) \rightarrow z$, we curry this into a function $g_{c1}(a) \rightarrow g_{c2}(b) \rightarrow g_{c2}(c) \rightarrow z$. This is done recursively. First, evaluation of the function $g_{c1}(a)$ returns a new function with the *b* and *c* arguments, $g'_{c1}(b, c)$. Then we can curry the returned two-argument function to create $g_{c2}(b) \rightarrow g_{c3}(c)$.

We can evaluate a complex curried function with `g_c(1)(2)(3)`. This formal syntax is bulky, so we use some syntactic sugar to reduce `g_c(1)(2)(3)` to something more palatable like `g(1, 2, 3)`.

Let's look at a concrete example in Python. For example, we have a function like the following one:

```
from pymonad import curry
@curry
def systolic_bp(bmi, age, gender_male, treatment):
    return (
        68.15+0.58*bmi+0.65*age+0.94*gender_male+6.44*treatment
    )
```

This is a simple, multiple-regression-based model for systolic blood pressure. This predicts blood pressure from **body mass index** (**BMI**), age, gender (a value of 1 means male), and history of previous treatment (a value of 1 means previously treated). For more information on the model and how it's derived, visit: `http://sphweb.bumc.bu.edu/otlt/MPH-Modules/BS/BS704_Multivariable/BS704_Multivariable7.html`.

We can use the `systolic_bp()` function with all four arguments, as follows:

```
>>> systolic_bp(25, 50, 1, 0)
116.09
>>> systolic_bp(25, 50, 0, 1)
121.59
```

A male person with a BMI of 25, age 50, and no previous treatment is predicted to have have a blood pressure near 116. The second example shows a similar woman with a history of treatment who will likely have a blood pressure of 121.

Because we've used the `@curry` decorator, we can create intermediate results that are similar to partially applied functions. Take a look at the following command snippet:

```
>>> treated = systolic_bp(25, 50, 0)
>>> treated(0)
115.15
>>> treated(1)
121.59
```

In the preceding case, we evaluated the `systolic_bp(25, 50, 0)` method to create a curried function and assigned this to the `treated` variable. The BMI, age, and gender values don't typically change for a given patient. We can now apply the new `treated` function to the remaining argument to get different blood pressure expectations based on patient history.

This is similar in some respects to the `functools.partial()` function. The important difference is that currying creates a function that can work in a variety of ways. The `functools.partial()` function creates a more specialized function that can only be used with the given set of bound values.

Here's an example of creating some additional curried functions:

```
>>> g_t= systolic_bp(25, 50)
>>> g_t(1, 0)
116.09
>>> g_t(0, 1)
121.59
```

This is a gender-based treatment function based on our initial model. We must provide gender and treatment values to get a final value from the model.

Using curried higher-order functions

While currying is simple to visualize using ordinary functions, the real value shows up when we apply currying to higher-order functions. In an ideal situation, the `functools.reduce()` function would be curryable so that we would be able to do this:

```
sum = reduce(operator.add)
prod = reduce(operator.mul)
```

This doesn't work, however. The built-in `reduce()` function isn't curryable by using the PyMonad library, so the above examples don't actually work. If we define our own `reduce()` function, however, we can then curry it as shown previously.

Here's an example of a home-brewed `reduce()` function that can be used as shown earlier:

```
from collections.abc import Sequence
from pymonad import curry

@curry
def myreduce(function, iterable_or_sequence):
    if isinstance(iterable_or_sequence, Sequence):
        iterator= iter(iterable_or_sequence)
    else:
        iterator= iterable_or_sequence
    s = next(iterator)
    for v in iterator:
        s = function(s,v)
    return s
```

The `myreduce()` function will behave like the built-in `reduce()` function. The `myreduce()` function works with an iterable or a sequence object. Given a sequence, we'll create an iterator; given an iterable object, we'll simply use it. We initialize the result with the first item in the iterator. We apply the function to the ongoing sum (or product) and each subsequent item.

It's also possible to wrap the built-in `reduce()` function to create a curryable version. That's only two lines of code; an exercise left for the reader.

Since the `myreduce()` function is a curried function, we can now use it to create functions based on our higher-order function, `myreduce()`:

```
>>> from operator import add
>>> sum = myreduce(add)
>>> sum([1,2,3])
6
>>> max = myreduce(lambda x,y: x if x > y else y)
>>> max([2,5,3])
5
```

We defined our own version of the `sum()` function using the curried reduce applied to the `add` operator. We also defined our own version of the default `max()` function using a `lambda` object that picks the larger of two values.

We can't easily create the more general form of the `max()` function this way, because currying is focused on positional parameters. Trying to use the `key=` keyword parameter adds too much complexity to make the technique work toward our overall goals of succinct and expressive functional programs.

To create a more generalized version of the `max()` function, we need to step outside the `key=` keyword parameter paradigm that functions such as `max()`, `min()`, and `sorted()` rely on. We would have to accept the higher-order function as the first argument in the same way as `filter()`, `map()`, and `reduce()` functions do. We could also create our own library of more consistent higher-order curried functions. These functions would rely exclusively on positional parameters. The higher-order function would be provided first so that our own curried `max(function, iterable)` method would follow the pattern set by the `map()`, `filter()`, and `functools.reduce()` functions.

Currying the hard way

We can create curried functions manually, without using the decorator from the PyMonad library; one way of doing this is shown in the function definition that follows:

```
def f(x, *args):
    def f1(y, *args):
        def f2(z):
            return (x+y)*z
        if args:
            return f2(*args)
        return f2
    if args:
        return f1(*args)
    return f1
```

This curries a function, $f(x, y, z) \rightarrow (x + y) * z$, into a function, `f(x)`, which returns a function. Conceptually, $f(x) \rightarrow f'(y, z)$. We then curried the intermediate function to create the `f1(y)` and `f2(z)` function, $f'(y, z) \rightarrow \left(f_2(y) \rightarrow f_3(z) \right)$.

When we evaluate the `f(x)` function, we'll get a new function, `f1`, as a result. If additional arguments are provided, those arguments are passed to the `f1` function for evaluation, either resulting in a final value or another function.

Clearly, this kind of manual expansion to implement currying is potentially error-prone. This isn't a practical approach to working with functions. It may, however, serve to illustrate what currying means and how it's implemented in Python.

Functional composition and the PyMonad * operator

One of the significant values of curried functions is the ability to combine them through functional composition. We looked at functional composition in `Chapter 5`, *Higher-Order Functions*, and `Chapter 11`, *Decorator Design Techniques*.

When we've created a curried function, we can more easily perform function composition to create a new, more complex curried function. In this case, the PyMonad package defines the `*` operator for composing two functions. To explain how this works, we'll define two curried functions that we can compose. First, we'll define a function that computes the product, and then we'll define a function that computes a specialized range of values.

Here's our first function, which computes the product:

```
import  operator
prod = myreduce(operator.mul)
```

This is based on our curried `myreduce()` function that was defined previously. It uses the `operator.mul()` function to compute a *times-reduction* of an iterable: we can call a product a times-reduce of a sequence.

Here's our second curried function that will produce a range of values:

```
@curry
def alt_range(n):
    if n == 0:
        return range(1, 2)  # Only the value [1]
    elif n % 2 == 0:
        return range(2, n+1, 2)
    else:
        return range(1, n+1, 2)
```

The result of the `alt_range()` function will be even values or odd values. It will have only odd values up to (and including) `n`, if `n` is odd. If `n` is even, it will have only even values up to `n`. The sequences are important for implementing the semifactorial or double factorial function, $n!!$.

Here's how we can combine the `prod()` and `alt_range()` functions into a new curried function:

```
>>> semi_fact = prod * alt_range
>>> semi_fact(9)
945
```

The PyMonad `*` operator here combines two functions into a composite function, named `semi_fact`. The `alt_range()` function is applied to the arguments. Then, the `prod()` function is applied to the results of the `alt_range` function.

Using the PyMonad `*` operator is equivalent to creating a new `lambda` object:

```
semi_fact = lambda x: prod(alt_range(x))
```

The composition of curried functions involves somewhat less syntax than creating a new `lambda` object.

Ideally, we would like to be able use functional composition and curried functions like this:

```
sumwhile = sum * takewhile(lambda x: x > 1E-7)
```

This could define a version of the `sum()` function that works with infinite sequences, stopping the generation of values when the threshold has been met. This doesn't actually work because the PyMonad library doesn't seem to handle infinite iterables as well as it handles the internal `List` objects.

Functors and applicative functors

The idea of a functor is a functional representation of a piece of simple data. A functor version of the number `3.14` is a function of zero arguments that returns this value. Consider the following example:

```
>>> pi = lambda: 3.14
>>> pi()
3.14
```

We created a zero-argument `lambda` object that returns a simple value.

When we apply a curried function to a functor, we're creating a new curried functor. This generalizes the idea of applying a function to an argument to get a value by using functions to represent the arguments, the values, and the functions themselves.

Once everything in our program is a function, then all processing is simply a variation of the theme of functional composition. The arguments and results of curried functions can be functors. At some point, we'll apply a `getValue()` method to a `functor` object to get a Python-friendly, simple type that we can use in uncurried code.

Since the programming is based on functional composition, no calculation needs to be done until we actually demand a value using the `getValue()` method. Instead of performing a lot of intermediate calculations, our program defines intermediate complex objects that can produce a value when requested. In principle, this composition can be optimized by a clever compiler or runtime system.

When we apply a function to a `functor` object, we're going to use a method similar to `map()` that is implemented as the `*` operator. We can think of the `function * functor` or `map(function, functor)` methods as a way to understand the role a functor plays in an expression.

In order to work politely with functions that have multiple arguments, we'll use the `&` operator to build composite functors. We'll often see the `functor & functor` method used to build a `functor` object from a pair of functors.

We can wrap Python simple types with a subclass of the `Maybe` functor. The `Maybe` functor is interesting, because it gives us a way to deal gracefully with missing data. The approach we used in `Chapter 11`, *Decorator Design Techniques*, was to decorate built-in functions to make them `None` aware. The approach taken by the PyMonad library is to decorate the data so that it gracefully declines being operated on.

There are two subclasses of the `Maybe` functor:

- `Nothing`
- `Just(some simple value)`

We use `Nothing` as a stand-in for the simple Python value of `None`. This is how we represent missing data. We use `Just(some simple value)` to wrap all other Python objects. These functors are function-like representations of constant values.

We can use a curried function with these `Maybe` objects to tolerate missing data gracefully. Here's a short example:

```
>>> x1 = systolic_bp * Just(25) & Just(50) & Just(1) & Just(0)
>>> x1.getValue()
116.09

>>> x2 = systolic_bp * Just(25) & Just(50) & Just(1) & Nothing
```

```
>>> x2.getValue() is None
True
```

The `*` operator is functional composition: we're composing the `systolic_bp()` function with an argument composite. The `&` operator builds a composite functor that can be passed as an argument to a curried function of multiple arguments.

This shows us that we get an answer instead of a `TypeError` exception. This can be very handy when working with large, complex datasets in which data could be missing or invalid. It's much nicer than having to decorate all of our functions to make them `None` aware.

This works nicely for curried functions. We can't operate on the `Maybe` functors in uncurried Python code as functors have very few methods.

We must use the `getValue()` method to extract the simple Python value for uncurried Python code.

Using the lazy List() functor

The `List()` functor can be confusing at first. It's extremely lazy, unlike Python's built-in `list` type. When we evaluate the built-in `list(range(10))` method, the `list()` function will evaluate the `range()` object to create a list with 10 items. The PyMonad `List()` functor, however, is too lazy to even do this evaluation.

Here's the comparison:

```
>>> list(range(10))
[0, 1, 2, 3, 4, 5, 6, 7, 8, 9]
>>> List(range(10))
[range(0, 10)]
```

The `List()` functor did not evaluate the `range()` object, it just preserved it without being evaluated. The `pymonad.List()` function is useful to collect functions without evaluating them.

The use of `range()` here can be a little confusing. The Python 3 `range()` object is also lazy. In this case, there are two tiers of laziness. The `pymonad.List()` will create items as needed. Each item in the `List` is a `range()` object which can be evaluated to produce a sequence of values.

We can evaluate the List functor later as required:

```
>>> x = List(range(10))
>>> x
[range(0, 10)]
>>> list(x[0])
[0, 1, 2, 3, 4, 5, 6, 7, 8, 9]
```

We created a lazy List object which contained a range() object. Then we extracted and evaluated a range() object at position 0 in that list.

A List object won't evaluate a generator function or range() object; it treats any iterable argument as a single iterator object. We can, however, use the * operator to expand the values of a generator or the range() object.

Note that there are several meanings for the * operator: it is the built-in mathematical times operator, the function composition operator defined by PyMonad, and the built-in modifier used when calling a function to bind a single sequence object as all of the positional parameters of a function. We're going to use the third meaning of the * operator to assign a sequence to multiple positional parameters.

Here's a curried version of the range() function. This has a lower bound of 1 instead of 0. It's handy for some mathematical work because it allows us to avoid the complexity of the positional arguments in the built-in range() function:

```
@curry
def range1n(n):
    if n == 0: return range(1, 2)  # Only the value 1
    return range(1, n+1)
```

We simply wrapped the built-in range() function to make it curryable by the PyMonad package.

Since a List object is a functor, we can map functions to the List object.

The function is applied to each item in the List object. Here's an example:

```
>>> fact= prod * range1n
>>> seq1 = List(*range(20))
>>> f1 = fact * seq1
>>> f1[:10]
[1, 1, 2, 6, 24, 120, 720, 5040, 40320, 362880]
```

We defined a composite function, `fact()`, which was built from the `prod()` and `range1n()` functions shown previously. This is the factorial function. We created a `List()` functor, `seq1`, which is a sequence of 20 values. We mapped the `fact()` function to the `seq1` functor, which created a sequence of factorial values, `f1`. We looked at the first 10 of these values earlier.

There is a similarity between the composition of functions and the composition of a function and a functor. Both `prod*range1n` and `fact*seq1` use functional composition: one combines things that are obviously functions, and the other combines a function and a functor.

Here's another little function that we'll use to extend this example:

```
@curry
def n21(n):
    return 2*n+1
```

This little `n21()` function does a simple computation. It's curried, however, so we can apply it to a functor such as a `List()` function. Here's the next part of the preceding example:

```
>>> semi_fact= prod * alt_range
>>> f2 = semi_fact * n21 * seq1
>>> f2[:10]
[1, 3, 15, 105, 945, 10395, 135135, 2027025, 34459425, 654729075]
```

We've defined a composite function from the `prod()` and `alt_range()` functions shown previously. The function `f2` is semifactorial or double factorial. The value of the function `f2` is built by mapping our small `n21()` function applied to the `seq1` sequence. This creates a new sequence. We then applied the `semi_fact` function to this new sequence to create a sequence of values that are parallels to the sequence of values.

We can now map the `/` operator to the `map()` and `operator.truediv` parallel functors:

```
>>> 2*sum(map(operator.truediv, f1, f2))
3.1415919276751456
```

The `map()` function will apply the given operator to both functors, yielding a sequence of fractions that we can add.

The `f1 & f2` method will create all combinations of values from the two `List` objects. This is an important feature of `List` objects: they readily enumerate all combinations allowing a simple algorithm to compute all alternatives and filter the alternatives for the proper subset. This is something we don't want; that's why we used the `map()` function instead of the `operator.truediv * f1 & f2` method.

We defined a fairly complex calculation using a few functional composition techniques and a functor class definition. Here's the full definition for this calculation:

$$\pi = \sum_{n=0}^{\infty} \frac{n!}{(2n+1)!!}$$

Ideally, we prefer not to use a fixed-sized `List` object. We'd prefer to have a lazy and potentially infinite sequence of integer values. We could then use a curried version of `sum()` and `takewhile()` functions to find the sum of values in the sequence until the values are too small to contribute to the result. This would require an even lazier version of the `List()` object that could work with the `itertools.counter()` function. We don't have this potentially infinite list in PyMonad 1.3; we're limited to a fixed-sized `List()` object.

Monad bind() function and the >> operator

The name of the PyMonad library comes from the functional programming concept of a `monad`, a function that has a strict order. The underlying assumption behind much functional programming is that functional evaluation is liberal: it can be optimized or rearranged as necessary. A monad provides an exception that imposes a strict left-to-right order.

Python, as we have seen, is strict. It doesn't *require* monads. We can, however, still apply the concept in places where it can help clarify a complex algorithm.

The technology for imposing strict evaluation is a binding between a monad and a function that will return a monad. A *flat* expression will become nested bindings that can't be reordered by an optimizing compiler. The `bind()` function is mapped to the >> operator, allowing us to write expressions like this:

```
Just(some file) >> read header >> read next >> read next
```

The preceding expression would be converted to the following:

```
bind(
    bind(
        bind(Just(some file), read header),
        read next),
    read next)
```

The `bind()` functions ensure that a strict left-to-right evaluation is imposed on this expression when it's evaluated. Also, note that the preceding expression is an example of functional composition. When we create a monad with the >> operator, we're creating a complex object that will be evaluated when we finally use the `getValue()` method.

The `Just()` subclass is required to create a simple monad compatible object that wraps a simple Python object.

The monad concept is central to expressing a strict evaluation order in a language that's heavily optimized and lenient. Python doesn't require a monad because it uses left-to-right strict evaluation. This makes the monad difficult to demonstrate because it doesn't really do something completely novel in a Python context. Indeed, the monad redundantly states the typical strict rules that Python follows.

In other languages, such as Haskell, a monad is crucial for file input and output where strict ordering is required. Python's imperative mode is much like a Haskell `do` block, which has an implicit Haskell >>= operator to force the statements to be evaluated in order. (PyMonad uses the `bind()` function and the >> operator for Haskell's >>= operation.)

Implementing simulation with monads

Monads are expected to pass through a kind of *pipeline*: a monad will be passed as an argument to a function and a similar monad will be returned as the value of the function. The functions must be designed to accept and return similar structures.

We'll look at a simple pipeline that can be used for simulation of a process. This kind of simulation may be a formal part of a Monte Carlo simulation. We'll take the phrase Monte Carlo simulation literally and simulate a casino dice game, Craps. This involves some stateful rules for a fairly complex simulation.

There is some strange gambling terminology involved, for which we apologize. Craps involves someone rolling the dice (a shooter) and additional bettors. The game has two phases of play:

- The first roll of the dice is called a *come out* roll. There are three conditions:
    - If the dice total is 7 or 11, the shooter wins. Anyone betting on the *pass* line will be paid off as a winner, and all other bets lose. The game is over, and the shooter can play again.
    - If the dice total is 2, 3, or 12, the shooter loses. Anyone betting on the *don't pass* line will win, and all other bets lose. The game is over, and the shooter must pass the dice to another shooter.
    - Any other total (that is, 4, 5, 6, 8, 9, or 10) establishes a *point*. The game changes state from the *come out* roll to the *point* roll. The game continues.
- Once a point is established, each *point* roll of the dice is evaluated with three conditions:
    - If the dice total is 7, the shooter loses. Almost all bets are losers except bets on the *don't pass* line and an obscure proposition bet. The game is over. Since the shooter lost, the dice is passed to another shooter.
    - If the dice totals the original point, the shooter wins. Anyone betting on the *pass* line will be paid off as a winner, and all other bets lose. The game is over, and the shooter can play again.
    - Any other total continues the game with no resolution. Some proposition bets may win or lose on these intermediate rolls.

The rules can be seen as requiring a state change. Alternatively, we can look at this as a sequence of operations rather than a state change. There's one function that must be used first. Another recursive function is used after that. In this way, this pairs-of-functions approach fits the monad design pattern nicely.

In a practical matter, a casino allows numerous fairly complex proposition bets during the game. We can evaluate those separately from the essential rules of the game. Many of those bets (the propositions, field bets, and buying a number) are made during the *point roll* phase of the game. We'll ignore this additional complexity and focus on the central game.

We'll need a source of random numbers:

```
import random
def rng():
    return (random.randint(1,6), random.randint(1,6))
```

The preceding function will generate a pair of dice for us.

Here's our expectations from the overall game:

```
def craps():
    outcome = (
        Just(("", 0, [])) >> come_out_roll(dice)
                          >> point_roll(dice)
    )
    print(outcome.getValue())
```

We create an initial monad, `Just(("",0, []))`, to define the essential type we're going to work with. A game will produce a three-tuple with the outcome text, the point value, and a sequence of rolls. At the start of each game, a default three tuple establishes the three-tuple type.

We pass this monad to two other functions. This will create a resulting monad, `outcome`, with the results of the game. We use the >> operator to connect the functions in the specific order they must be executed. In a language with an optimizing compiler, this will prevent the expression from being rearranged.

We get the value of the monad at the end using the `getValue()` method. Since the monad objects are lazy, this request is what triggers the evaluation of the various monads to create the required output.

The `come_out_roll()` function has the `rng()` function curried as the first argument. The monad will become the second argument to this function. The `come_out_roll()` function can roll the dice and apply the *come out* rule to determine if we have a win, a loss, or a point.

The `point_roll()` function also has the `rng()` function curried as the first argument. The monad will become the second argument. The `point_roll()` function can then roll the dice to see if the bet is resolved. If the bet is unresolved, this function will operate recursively to continue looking for a resolution.

The `come_out_roll()` function looks like this:

```
@curry
def come_out_roll(dice, status):
    d = dice()
    if sum(d) in (7, 11):
        return Just(("win", sum(d), [d]))
    elif sum(d) in (2, 3, 12):
        return Just(("lose", sum(d), [d]))
    else:
        return Just(("point", sum(d), [d]))
```

We roll the dice once to determine if the first roll will win, lose, or establish the point. We return an appropriate monad value that includes the outcome, a point value, and the roll of the dice. The point values for an immediate win and immediate loss aren't really meaningful. We could sensibly return a `0` value here, since no point was really established.

The `point_roll()` function looks like this:

```
@curry
def point_roll(dice, status):
    prev, point, so_far = status
    if prev != "point":
        return Just(status)
    d = dice()
    if sum(d) == 7:
        return Just(("craps", point, so_far+[d]))
    elif sum(d) == point:
        return Just(("win", point, so_far+[d]))
    else:
        return (
            Just(("point", point, so_far+[d]))
            >> point_roll(dice)
        )
```

We decomposed the `status` monad into the three individual values of the tuple. We could have used small `lambda` objects to extract the first, second, and third values. We could also have used the `operator.itemgetter()` function to extract the tuples' items. Instead, we used multiple assignment.

If a point was not established, the previous state will be *win* or *lose*. The game was resolved in a single throw, and this function simply returns the `status` monad.

If a point was established, the state will be *point*. The dice is rolled and rules applied to this new roll. If roll is 7, the game is a lost and a final monad is returned. If the roll is the point, the game is won and the appropriate monad is returned. Otherwise, a slightly revised monad is passed to the `point_roll()` function. The revised `status` monad includes this roll in the history of rolls.

A typical output looks like this:

```
>>> craps()
('craps', 5, [(2, 3), (1, 3), (1, 5), (1, 6)])
```

The final monad has a string that shows the outcome. It has the point that was established and the sequence of dice rolls. Each outcome has a specific payout that we can use to determine the overall fluctuation in the bettor's stake.

We can use simulation to examine different betting strategies. We might be searching for a way to defeat any house edge built into the game.

There's some small asymmetry in the basic rules of the game. Having 11 as an immediate winner is balanced by having 3 as an immediate loser. The fact that 2 and 12 are also losers is the basis of the house's edge of 5.5 percent (*1/18 = 5.5*) in this game. The idea is to determine which of the additional betting opportunities will dilute this edge.

A great deal of clever Monte Carlo simulation can be built with a few simple, functional programming design techniques. The monad, in particular, can help to structure these kinds of calculations when there are complex orders or internal states.

Additional PyMonad features

One of the other features of PyMonad is the confusingly named **monoid**. This comes directly from mathematics and it refers to a group of data elements that have an operator and an identity element, and the group is closed with respect to that operator. When we think of natural numbers, the `add` operator, and an identity element `0`, this is a proper monoid. For positive integers, with an operator `*`, and an identity value of `1`, we also have a monoid; strings using `|` as an operator and an empty string as an identity element also qualify.

PyMonad includes a number of predefined monoid classes. We can extend this to add our own `monoid` class. The intent is to limit a compiler to certain kinds of optimization. We can also use the monoid class to create data structures which accumulate a complex value, perhaps including a history of previous operations.

Much of this provides insight into functional programming. To paraphrase the documentation, this is an easy way to learn about functional programming in, perhaps, a slightly more forgiving environment. Rather than learning an entire language and toolset to compile and run functional programs, we can just experiment with interactive Python.

Pragmatically, we don't need too many of these features because Python is already stateful and offers strict evaluation of expressions. There's no practical reason to introduce stateful objects in Python, or strictly ordered evaluation. We can write useful programs in Python by mixing functional concepts with Python's imperative implementation. For that reason, we won't delve more deeply into PyMonad.

Summary

In this chapter, we looked at how we can use the PyMonad library to express some functional programming concepts directly in Python. The module contains many important functional programming techniques.

We looked at the idea of currying, a function that allows combinations of arguments to be applied to create new functions. Currying a function also allows us to use functional composition to create more complex functions from simpler pieces. We looked at functors that wrap simple data objects to make them into functions which can also be used with functional composition.

Monads are a way to impose a strict evaluation order when working with an optimizing compiler and lazy evaluation rules. In Python, we don't have a good use case for monads, because Python is an imperative programming language under the hood. In some cases, imperative Python may be more expressive and succinct than a monad construction.

In the next chapter, we'll look at how we can apply functional programming techniques to build web service applications. The idea of HTTP can be summarized as `response = httpd(request)`. Ideally, HTTP is stateless, making it a perfect match for functional design. Using cookies is analogous to providing a response value which is expected as an argument to a later request.

15
A Functional Approach to Web Services

We'll step away from the topic of *Exploratory Data Analysis* to look at web servers and web services. A web server is, to an extent, a cascade of functions. We can apply a number of functional design patterns to the problem of presenting web content. Our goal is to look at ways in which we can approach **Representational State Transfer** (**REST**). We want to build RESTful web services using functional design patterns.

We don't need to invent yet another Python web framework. Nor do we want to select from among the available frameworks. There are many web frameworks available in Python, each with a distinct set of features and advantages.

The intent of this chapter is to present some principles that can be applied to most of the available frameworks. This will let us leverage functional design patterns for presenting web content.

When we look at extremely large or complex datasets, we might want a web service that supports subsetting or searching. We might also want a website that can download subsets in a variety of formats. In this case, we might need to use functional designs to create RESTful web services to support these more sophisticated requirements.

Interactive web applications often rely on stateful sessions to make the site easier for people to use. A user's session information is updated with data provided through HTML forms, fetched from databases, or recalled from caches of previous interactions. Because the stateful data must be fetched as part of each transaction, it becomes more like an input parameter or result value. This can lead to functional-style programming even in the presence of cookies and database updates.

In this chapter, we'll look at several topics:

- The general idea of the HTTP request and response model.
- The **Web Services Gateway Interface** (**WSGI**) standard that Python applications use.
- Leveraging WSGI, where it's possible to define web services as functions. This fits with the HTTP idea of a stateless server.
- We'll also look at ways to authorize client applications to make use of a web service.

The HTTP request-response model

The HTTP protocol is nearly stateless: a user agent (or browser) makes a request and the server provides a response. For services that don't involve cookies, a client application can take a functional view of the protocol. We can build a client using the `http.client` or `urllib` library. An HTTP user agent essentially executes something similar to the following:

```
import urllib.request
def urllib_demo(url):
    with urllib.request.urlopen(url) as response:
        print(response.read())

urllib_demo("http://slott-softwarearchitect.blogspot.com")
```

A program like **wget** or **curl** does this kind of processing using a URL supplied as a command-line argument. A browser does this in response to the user pointing and clicking; the URL is taken from the user's actions, often the action of clicking on linked text or images.

The practical considerations of user-experience (UX) design, however, lead to some implementation details that are stateful. When a client is obliged to track cookies, it becomes stateful. A response header will provide cookies, and subsequent requests must return the cookies to the server. We'll look at this in more detail later.

An HTTP response can include a status code that requires additional actions on the part of the user agent. Many status codes in the 300-399 range indicate that the requested resource has been moved. The user agent is then required to save details from the `Location` header and request a new URL. The `401` status code indicates that authentication is required; the user agent must make another request using the `Authorization` header that contains credentials for access to the server. The `urllib` library implementation handles this stateful client processing. The `http.client` library doesn't automatically follow `3xx` redirect status codes.

The techniques for a user agent to handle `3xx` and `401` codes can be handled through a simple recursion. If the status doesn't indicate a redirection, this is the base case, and the function has a result. If redirection is required, the function can be called recursively with the redirected address.

Looking at the other side of the protocol, a static content server can be stateless. We can use the `http.server` library for this, as follows:

```
from http.server import HTTPServer, def server_demo():
httpd = HTTPServer(
    ('localhost', 8080), SimpleHTTPRequestHandler)
while True:
    httpd.handle_request()
httpd.shutdown()
```

We created a server object, and assigned it to the `httpd` variable. We provided the address and port number to which we'll listen for connection requests. The TCP/IP protocol will spawn a connection on a separate port. The HTTP protocol will read the request from this other port and create an instance of the handler.

In this example, we provided `SimpleHTTPRequestHandler` as the class to instantiate with each request. This class must implement a minimal interface, which will send headers and then send the body of the response to the client. This particular class will serve files from the local directory. If we wish to customize this, we can create a subclass, which implements methods such as `do_GET()` and `do_POST()` to alter the behavior.

Often, we use the `serve_forever()` method, instead of writing our own loop. We've shown the loop here to clarify that the server must, generally, be crashed if we need to stop it.

Injecting state through cookies

The addition of cookies changes the overall relationship between a client and server to become stateful. Interestingly, it involves no change to the HTTP protocol itself. The state information is communicated through headers on the request and the reply. The server will send cookies to the user agent in response headers. The user agent will save and reply with cookies in request headers.

The user agent or browser is required to retain a cache of cookie values, provided as part of a response, and include appropriate cookies in subsequent requests. The web server will look for cookies in the request header and provide updated cookies in the response header. The effect is to make the web server stateless; the state changes happen only in the client. A server sees cookies as additional arguments in a request and provides additional details in a response.

Cookies can contain anything. They are often encrypted to avoid exposing web server details to other applications running on the client computer. Transmitting large cookies can slow down processing. Optimizing this kind of state processing is best handled by an existing framework. We'll ignore details of cookies and session management.

The concept of a *session* is a feature of a web server, not the HTTP protocol. A session is commonly defined as a series of requests with the same cookie. When an initial request is made, no cookie is available, and a new session cookie is created. Every subsequent request will include the cookie. A logged-in user will have additional details in their session cookie. A session can last as long as the server is willing to accept the cookie: a cookie could be valid forever, or expire after a few minutes.

The REST approach to web services does not rely on sessions or cookies. Each REST request is distinct. This makes it less *user-friendly* than an interactive site that uses cookies to simplify a user's interactions. We'll focus on RESTful web services because they fit the functional design patterns very closely.

One consequence of sessionless REST processes is each individual REST request is separately authenticated. If authentication is required, it means REST traffic must use **Secured Socket Layer** (**SSL**) protocols; the `https` scheme can be used to transmit credentials securely from client to server.

Considering a server with a functional design

One core idea behind HTTP is that the server's response is a function of the request. Conceptually, a web service should have a top-level implementation that can be summarized as follows:

```
response = httpd(request)
```

However, this is impractical. It turns out that an HTTP request isn't a simple, monolithic data structure. It has some required parts and some optional parts. A request may have headers, a method and a path, and there may be attachments. The attachments may include forms or uploaded files, or both.

To make things more complex, a browser's form data can be sent as a query string in the path of a `GET` request. Alternatively, it can be sent as an attachment to a `POST` request. While there's a possibility for confusion, most web application frameworks will create HTML form tags that provide their data through a `"method=POST"` argument in the `<form>` tag; the form data will then be included in the request as an attachment.

Looking more deeply into the functional view

Both HTTP responses and requests have headers separate from the body. The request can also have some attached form data or other uploads. Therefore, we can think of a web server like this:

```
headers, content = httpd(headers, request, [form or uploads])
```

The request headers may include cookie values, which can be seen as adding yet more arguments. Additionally, a web server is often dependent on the OS environment in which it's running. This OS environment data can be considered as yet more arguments being provided as part of the request.

There's a large but reasonably well-defined spectrum of content. The **Multipurpose Internet Mail Extension (MIME)** types define the kinds of content that a web service might return. This can include plain text, HTML, JSON, XML, or any of the wide variety of non-text media that a website might serve.

As we look more closely at the processing required to build a response to an HTTP request, we'll see some common features that we'd like to reuse. This idea of reusable elements is what leads to the creation of web service frameworks that fill a spectrum from simple to sophisticated. The ways that functional designs allow us to reuse functions indicate that the functional approach can help building web services.

We'll look at functional design of web services by examining how we can create a pipeline of the various elements of a service response. We'll do this by nesting the functions for request processing so that inner elements are free from the generic overheads, which are provided by outer elements. This also allows the outer elements to act as filters: invalid requests can yield error responses, allowing the inner function to focus narrowly on the application processing.

Nesting the services

We can look at web-request handling as a number of nested contexts. An outer context, for example, might cover session management: examining the request to determine if this is another request in an existing session or a new session. An inner context might provide tokens used for form processing that can detect **Cross-Site Request Forgeries** (**CSRF**). Another context might handle user authentication within a session.

A conceptual view of the functions explained previously is something like this:

```
response = content(
    authentication(
        csrf(
            session(headers, request, forms)
        )
    )
)
```

The idea here is that each function can build on the results of the previous function. Each function either enriches the request or rejects it because it's invalid. The `session()` function, for example, can use headers to determine if this is an existing session or a new session. The `csrf()` function will examine form input to ensure that proper tokens were used. The CSRF handling requires a valid session. The `authentication()` function can return an error response for a session that lacks valid credentials; it can enrich the request with user information when valid credentials are present.

The `content()` function is free from worrying about sessions, forgeries, and non-authenticated users. It can focus on parsing the path to determine what kind of content should be provided. In a more complex application, the `content()` function may include a rather complex mapping from path elements to the functions that determine the appropriate content.

The nested function view, however, still isn't quite right. The problem is that each nested context may also need to tweak the response instead of, or in addition to, tweaking the request.

We really want something more like this:

```
def session(headers, request, forms):
    pre-process: determine session
    content = csrf(headers, request, forms)
     post-processes the content
     return the content

def csrf(headers, request, forms):
    pre-process: validate csrf tokens
    content = authenticate(headers, request, forms)
    post-processes the content
    return the content
```

This concept points toward a functional design for creating web content through a nested collection of functions that provide enriched input or enriched output, or both. With a little bit of cleverness, we should be able to define a simple, standard interface that various functions can use. Once we've standardized an interface, we can combine functions in different ways and add features. We should be able to meet our functional programming objectives of having succinct and expressive programs that provide web content.

The WSGI standard

The **Web Server Gateway Interface** (**WSGI**) defines a relatively simple, standardized design pattern for creating a response to a web request. This is a common framework for most Python-based web servers. A great deal of information is present at the following link: `http://wsgi.readthedocs.org/en/latest/`.

Some important background of WSGI can be found at `https://www.python.org/dev/peps/pep-0333`.

The Python library's `wsgiref` package includes a reference implementation of WSGI. Each WSGI *application* has the same interface, as shown here:

```
def some_app(environ, start_response):
    return content
```

The `environ` parameter is a dictionary that contains all of the arguments of the request in a single, uniform structure. The headers, the request method, the path, and any attachments for forms or file uploads will all be in the environment. In addition to this, the OS-level context is also provided along with a few items that are part of WSGI request handling.

The `start_response` parameter is a function that must be used to send the status and headers of a response. The portion of a WSGI server that has the final responsibility for building the response will use the given `start_response()` function, and will also build the response document as the return value.

The response returned from a WSGI application is a sequence of strings or string-like file wrappers that will be returned to the user agent. If an HTML template tool is used, then the sequence may have a single item. In some cases, such as the **Jinja2** templates, the template can be rendered lazily as a sequence of text chunks. This allows a server to interleave template filling with downloading to the user agent.

We can use the following type hints for a WSGI application:

```
from typing import (
    Dict, Callable, List, Tuple, Iterator, Union, Optional
)
from mypy_extensions import DefaultArg

SR_Func =  Callable[
    [str, List[Tuple[str, str]], DefaultArg(Tuple)], None]

def static_app(
    environ: Dict,
    start_response: SR_Func
  ) -> Union[Iterator[bytes], List[bytes]]:
```

The `SR_Func` type definition is the signature for the `start_response` function. Note that the function has an optional argument, requiring a function from the `mypy_extensions` module to define this feature.

An overall WSGI function, `static_app()`, requires the environment and the `start_response()` function. The result is either a sequence of bytes or an iterator over bytes. The union of return types from the `static_app()` function could be expanded to include `BinaryIO` and `List[BinaryIO]`, but they're not used by any of the examples in this chapter.

As of the publication date, the `wsgiref` package does not have a complete set of type definitions. Specifically, the `wsgiref.simple_server` module lacks appropriate stub definitions, and will result in warnings from **mypy**.

Each WSGI application is designed as a collection of functions. The collection can be viewed as nested functions or as a chain of transformations. Each application in the chain will either return an error or will hand the request to another application that will determine the final result.

Often, the URL path is used to determine which of many alternative applications, will be used. This will lead to a tree of WSGI applications which may share common components.

Here's a very simple routing application that takes the first element of the URL path and uses this to locate another WSGI application that provides content:

```
SCRIPT_MAP = {
    "demo": demo_app,
    "static": static_app,
    "index.html": welcome_app,
}
def routing(environ, start_response):
    top_level = wsgiref.util.shift_path_info(environ)
    app = SCRIPT_MAP.get(top_level, welcome_app)
    content = app(environ, start_response)
    return content
```

This applications will use the `wsgiref.util.shift_path_info()` function to tweak the environment. The change is a *head/tail split* on the items in the request path, available in the `environ['PATH_INFO']` dictionary. The head of the path up to the first `"/"`-will be moved into the `SCRIPT_NAME` item in the environment; the `PATH_INFO` item will be updated to have the tail of the path. The returned value will also be the head of the path, the same value as `environ['SCRIPT_NAME']`. In the case where there's no path to parse, the return value is `None` and no environment updates are made.

The `routing()` function uses the first item on the path to locate an application in the `SCRIPT_MAP` dictionary. We use the `welcome_app` as a default in case the requested path doesn't fit the mapping. This seems a little better than an HTTP `404 NOT FOUND` error.

This WSGI application is a function that chooses between a number of other WSGI functions. Note that the routing function doesn't return a function; it provides the modified environment to the selected WSGI application. This is the typical design pattern for handing off the work from one function to another.

It's easy to see how a framework could generalize the path-matching process, using regular expressions. We can imagine configuring the `routing()` function with a sequence of **Regular Expressions (REs)** and WSGI applications, instead of a mapping from a string to the WSGI application. The enhanced `routing()` function application would evaluate each RE looking for a match. In the case of a match, any `match.groups()` function could be used to update the environment before calling the requested application.

Throwing exceptions during WSGI processing

One central feature of WSGI applications is that each stage along the chain is responsible for filtering the requests. The idea is to reject faulty requests as early in the processing as possible. Python's exception handling makes this particularly simple.

We can define a WSGI application that provides static content as follows:

```
def static_app(
        environ: Dict,
        start_response: SR_Func
    ) -> Union[Iterator[bytes], List[bytes]]:
    log = environ['wsgi.errors']
    try:
        print(f"CWD={Path.cwd()}", file=log)
        static_path = Path.cwd()/environ['PATH_INFO'][1:]
        with static_path.open() as static_file:
            content = static_file.read().encode("utf-8")
            headers = [
                ("Content-Type", 'text/plain;charset="utf-8"'),
                ("Content-Length", str(len(content))),
            ]
            start_response('200 OK', headers)
            return [content]
    except IsADirectoryError as e:
        return index_app(environ, start_response)
    except FileNotFoundError as e:
```

```
        start_response('404 NOT FOUND', [])
        return [
            f"Not Found {static_path}\n{e!r}".encode("utf-8")
        ]
```

This application creates a `Path` object from the current working directory and an element of the path provided as part of the requested URL. The path information is part of the WSGI environment, in an item with the `'PATH_INFO'` key. Because of the way the path is parsed, it will have a leading `"/"`, which we discard, by using `environ['PATH_INFO][1:]`.

This application tries to open the requested path as a text file. There are two common problems, both of which are handled as exceptions:

- If the file is a directory, we'll use a different application, `index_app`, to present directory contents
- If the file is simply not found, we'll return an `HTTP 404 NOT FOUND` response

Any other exceptions raised by this WSGI application will not be caught. The application that invoked this application should be designed with some generic error-response capability. If the application doesn't handle the exceptions, a generic WSGI failure response will be used.

Our processing involves a strict ordering of operations. We must read the entire file so that we can create a proper HTTP `Content-Length` header.

There are two variations here on handling an exception. In one case, another application was invoked. If additional information needs to be provided to the other application, the environment must be updated with the required information. This is how a standardized error page can be built for a complex website.

The other case invoked the `start_response()` function and returned an error result. This is appropriate for unique, localized behavior. The final content is provided as bytes. This means that the Python strings must be properly encoded, and we must provide the encoding information to the user agent. Even the error message `repr(e)` is properly encoded before being downloaded.

Pragmatic WSGI applications

The intent of the WSGI standard is not to define a complete web framework; the intent is to define a minimum set of standards that allows flexible interoperability of web-related processing. A framework can take a wildly different approach to providing web services. The outermost interface should be compatible with WSGI, so that it can be used in a variety of contexts.

Web servers such as Apache **httpd** and **Nginx** have adapters to provide a WSGI-compatible interface from the web server to Python applications. For more information on WSGI implementations, visit: `https://wiki.python.org/moin/WSGIImplementations`.

Embedding our applications in a larger server allows us to have a tidy separation of concerns. We can use Apache httpd to serve completely static content, such as `.css`, `.js`, and image files. For HTML pages, though, a server like NGINX can use the `uwsgi` module to hand off requests to a separate Python process, which handles only the interesting HTML portions of the web content.

Separating the static content from the dynamic content means that we must either create a separate media server, or define our website to have two sets of paths. If we take the second approach, some paths will have the completely static content and can be handled by Nginx. Other paths will have dynamic content, which will be handled by Python.

When working with WSGI functions, it's important to note that we can't modify or extend the WSGI interface in any way. The point is complete compatibility with the externally visible layer of an application. Internal structures and processing do not have to conform to the WSGI standard. The external interface must follow these rules without exception.

A consequence of the WSGI definition is that the `environ` dictionary is often updated with additional configuration parameters. In this way, some WSGI applications can serve as gateways to enrich the environment with information extracted from cookies, configuration files, or databases.

Defining web services as functions

We'll look at a RESTful web service, which can slice and dice a source of data and provide downloads as JSON, XML, or CSV files. We'll provide an overall WSGI-compatible wrapper. The functions that do the real work of the application won't be narrowly constrained to fit the WSGI standard.

We'll use a simple dataset with four subcollections: the Anscombe Quartet. We looked at ways to read and parse this data in `Chapter 3`, *Functions, Iterators, and Generators*. It's a small set of data but it can be used to show the principles of a RESTful web service.

We'll split our application into two tiers: a web tier, which will be a simple WSGI application, and data service tier, which will be more typical functional programming. We'll look at the web tier first so that we can focus on a functional approach to provide meaningful results.

We need to provide two pieces of information to the web service:

- The quartet that we want: this is a slice and dice operation. The idea is to slice up the information by filtering and extracting meaningful subsets.
- The output format we want.

The data selection is commonly done through the request path. We can request `/anscombe/I/` or `/anscombe/II/` to pick specific datasets from the quartet. The idea is that a URL defines a resource, and there's no good reason for the URL to ever change. In this case, the dataset selectors aren't dependent on dates or some organizational approval status, or other external factors. The URL is timeless and absolute.

The output format is not a first-class part of the URL. It's just a serialization format, not the data itself. In some cases, the format is requested through the HTTP `Accept` header. This is hard to use from a browser, but easy to use from an application using a RESTful API. When extracting data from the browser, a query string is commonly used to specify the output format. We'll use the `?form=json` method at the end of the path to specify the JSON output format.

A URL we can use will look like this:

```
http://localhost:8080/anscombe/III/?form=csv
```

This would request a CSV download of the third dataset.

Creating the WSGI application

First, we'll use a simple URL pattern-matching expression to define the one and only routing in our application. In a larger or more complex application, we might have more than one such pattern:

```
import re
path_pat= re.compile(r"^/anscombe/(?P<dataset>.*?)/?$")
```

This pattern allows us to define an overall *script* in the WSGI sense at the top level of the path. In this case, the script is `anscombe`. We'll take the next level of the path as a dataset to select from the Anscombe Quartet. The dataset value should be one of `I`, `II`, `III`, or `IV`.

We used a named parameter for the selection criteria. In many cases, RESTful APIs are described using a syntax, as follows:

```
/anscombe/{dataset}/
```

We translated this idealized pattern into a proper, regular expression, and preserved the name of the dataset selector in the path.

Here are some example URL paths that demonstrate how this pattern works:

```
>>> m1 = path_pat.match( "/anscombe/I" )
>>> m1.groupdict()
{'dataset': 'I'}
>>> m2 = path_pat.match( "/anscombe/II/" )
>>> m2.groupdict()
{'dataset': 'II'}
>>> m3 = path_pat.match( "/anscombe/" )
>>> m3.groupdict()
{'dataset': ''}
```

Each of these examples shows the details parsed from the URL path. When a specific series is named, this is located in the path. When no series is named, then an empty string is found by the pattern.

Here's the overall WSGI application:

```
import traceback
import urllib.parse
def anscombe_app(
        environ: Dict, start_response: SR_Func
    ) -> Iterable[bytes]:
    log = environ['wsgi.errors']
    try:
        match = path_pat.match(environ['PATH_INFO'])
        set_id = match.group('dataset').upper()
        query = urllib.parse.parse_qs(environ['QUERY_STRING'])
        print(environ['PATH_INFO'], environ['QUERY_STRING'],
              match.groupdict(), file=log)

        dataset = anscombe_filter(set_id, raw_data())
        content_bytes, mime = serialize(
            query['form'][0], set_id, dataset)
```

```
        headers = [
            ('Content-Type', mime),
            ('Content-Length', str(len(content_bytes))),
        ]
        start_response("200 OK", headers)
        return [content_bytes]
    except Exception as e:  # pylint: disable=broad-except
        traceback.print_exc(file=log)
        tb = traceback.format_exc()
        content = error_page.substitute(
            title="Error", message=repr(e), traceback=tb)
        content_bytes = content.encode("utf-8")
        headers = [
            ('Content-Type', "text/html"),
            ('Content-Length', str(len(content_bytes))),
        ]
        start_response("404 NOT FOUND", headers)
        return [content_bytes]
```

This application will extract two pieces of information from the request: the `PATH_INFO` and the `QUERY_STRING` keys in the environment dictionary. The `PATH_INFO` request will define which set to extract. The `QUERY_STRING` request will specify an output format.

It's important to note that query strings can be quite complex. Rather than assume it is simply a string like `?form=json`, we've used the `urllib.parse` module to properly locate all of the name-value pairs in the query string. The value with the `'form'` key in the dictionary extracted from the query string can be found in `query['form'][0]`. This should be one of the defined formats. If it isn't, an exception will be raised, and an error page displayed.

After locating the path and query string, the application processing is highlighted in bold. These two statements rely on three functions to gather, filter, and serialize the results:

- The `raw_data()` function reads the raw data from a file. The result is a dictionary with lists of `Pair` objects.
- The `anscombe_filter()` function accepts a selection string and the dictionary of raw data and returns a single list of `Pair` objects.
- The list of pairs is then serialized into bytes by the `serialize()` function. The serializer is expected to produce byte's, which can then be packaged with an appropriate header, and returned.

We elected to produce an HTTP `Content-Length` header as part of the result. This header isn't required, but it's polite for large downloads. Because we decided to emit this header, we are forced to create a bytes object with the serialization of the data so we can count the bytes.

If we elected to omit the `Content-Length` header, we could change the structure of this application dramatically. Each serializer could be changed to a generator function, which would yield bytes as they are produced. For large datasets, this can be a helpful optimization. For the user watching a download, however, it might not be so pleasant because the browser can't display how much of the download is complete.

A common optimization is to break the transaction into two parts. The first part computes the result and places a file into a `Downloads` directory. The response is a `302 FOUND` with a `Location` header that identifies the file to download. Generally, most clients will then request the file based on this initial response. The file can be downloaded by Apache **httpd** or **Nginx** without involving the Python application.

For this example, all errors are treated as a `404 NOT FOUND` error. This could be misleading, since a number of individual things might go wrong. More sophisticated error handling could give more `try:/except:` blocks to provide more informative feedback.

For debugging purposes, we've provided a Python stack trace in the resulting web page. Outside the context of debugging, this is a very bad idea. Feedback from an API should be just enough to fix the request, and nothing more. A stack trace provides too much information to potentially malicious users.

Getting raw data

The `raw_data()` function is similar to the example shown in `Chapter 3`, *Functions, Iterators, and Generators*. We included some important changes. Here's what we're using for this application:

```
from Chapter_3.ch03_ex5 import (
    series, head_map_filter, row_iter)
from typing import (
    NamedTuple, Callable, List, Tuple, Iterable, Dict, Any)

RawPairIter = Iterable[Tuple[float, float]]

class Pair(NamedTuple):
    x: float
    y: float
```

```
pairs: Callable[[RawPairIter], List[Pair]] \
    = lambda source: list(Pair(*row) for row in source)

def raw_data() -> Dict[str, List[Pair]]:
    with open("Anscombe.txt") as source:
        data = tuple(head_map_filter(row_iter(source)))
        mapping = {
            id_str: pairs(series(id_num, data))
            for id_num, id_str in enumerate(
                ['I', 'II', 'III', 'IV'])
        }
    return mapping
```

The `raw_data()` function opens the local data file, and applies the `row_iter()` function to return each line of the file parsed into a row of separate items. We applied the `head_map_filter()` function to remove the heading from the file. The result created a tuple-of-list structure, which is assigned the variable `data`. This handles parsing the input into a structure that's useful. The resulting structure is an instance of the `Pair` subclass of the `NamedTuple` class, with two fields that have `float` as their type hints.

We used a dictionary comprehension to build the mapping from `id_str` to pairs assembled from the results of the `series()` function. The `series()` function extracts (x, y) pairs from the input document. In the document, each series is in two adjacent columns. The series named `I` is in columns zero and one; the `series()` function extracts the relevant column pairs.

The `pairs()` function is created as a `lambda` object because it's a small generator function with a single parameter. This function builds the desired `NamedTuple` objects from the sequence of anonymous tuples created by the `series()` function.

Since the output from the `raw_data()` function is a mapping, we can do something like the following example to pick a specific series by name:

```
>>> raw_data()['I']
[Pair(x=10.0, y=8.04), Pair(x=8.0, y=6.95), ...
```

Given a key, for example, `'I'`, the series is a list of `Pair` objects that have the x, y values for each item in the series.

Applying a filter

In this application, we're using a very simple filter. The entire filter process is embodied in the following function:

```
def anscombe_filter(
        set_id: str, raw_data_map: Dict[str, List[Pair]]
    ) -> List[Pair]:
    return raw_data_map[set_id]
```

We made this trivial expression into a function for three reasons:

- The functional notation is slightly more consistent and a bit more flexible than the subscript expression
- We can easily expand the filtering to do more
- We can include separate unit tests in the docstring for this function

While a simple `lambda` would work, it wouldn't be quite as convenient to test.

For error handling, we've done exactly nothing. We've focused on what's sometimes called the *happy path*: an ideal sequence of events. Any problems that arise in this function will raise an exception. The WSGI wrapper function should catch all exceptions and return an appropriate status message and error response content.

For example, it's possible that the `set_id` method will be wrong in some way. Rather than obsess over all the ways it could be wrong, we'll simply allow Python to throw an exception. Indeed, this function follows the Python advice that, *it's better to seek forgiveness than to ask permission*. This advice is materialized in code by avoiding *permission-seeking*: there are no preparatory `if` statements that seek to qualify the arguments as valid. There is only *forgiveness* handling: an exception will be raised and handled in the WSGI wrapper. This essential advice applies to the preceding raw data and the serialization that we will see now.

Serializing the results

Serialization is the conversion of Python data into a stream of bytes, suitable for transmission. Each format is best described by a simple function that serializes just that one format. A top-level generic serializer can then pick from a list of specific serializers. The picking of serializers leads to the following collection of functions:

```
Serializer = Callable[[str, List[Pair]], bytes]
SERIALIZERS: Dict[str, Tuple[str, Serializer]]= {
```

```
        'xml': ('application/xml', serialize_xml),
        'html': ('text/html', serialize_html),
        'json': ('application/json', serialize_json),
        'csv': ('text/csv', serialize_csv),
}

def serialize(
            format: str, title: str, data: List[Pair]
        ) -> Tuple[bytes, str]:
        mime, function = SERIALIZERS.get(
            format.lower(), ('text/html', serialize_html))
        return function(title, data), mime
```

The overall `serialize()` function locates a specific serializer in the `SERIALIZERS` dictionary, which maps a format name to a two-tuple. The tuple has a MIME type that must be used in the response to characterize the results. The tuple also has a function based on the `Serializer` type hint. This function will transform a name and a list of `Pair` objects into bytes that will be downloaded.

The `serialize()` function doesn't do any data transformation. It merely maps a name to a function that does the hard work of transformation. Returning a function permits the overall application to manage the details of memory or file-system serialization. Serializing to the file system, while slow, permits larger files to be handled.

We'll look at the individual serializers below. The serializers fall into two groups: those that produce strings and those that produce bytes. A serializer that produces a string will need to have the string encoded as bytes for download. A serializer that produces bytes doesn't need any further work.

For the serializers, which produce strings, we can use function composition with a standardized convert-to-bytes function. Here's a decorator that can standardize the conversion to bytes:

```
from typing import Callable, TypeVar, Any, cast

from functools import wraps
def to_bytes(
            function: Callable[..., str]
        ) -> Callable[..., bytes]:
        @wraps(function)
        def decorated(*args, **kw):
            text = function(*args, **kw)
            return text.encode("utf-8")
        return cast(Callable[..., bytes], decorated)
```

We've created a small decorator named `@to_bytes`. This will evaluate the given function and then encode the results using UTF-8 to get bytes. Note that the decorator changes the decorated function from having a return type of `str` to a return type of `bytes`. We haven't formally declared parameters for the decorated function, and used `...` instead of the details. We'll show how this is used with JSON, CSV, and HTML serializers. The XML serializer produces bytes directly and doesn't need to be composed with this additional function.

We could also do the functional composition in the initialization of the `serializers` mapping. Instead of decorating the function definition, we could decorate the reference to the function object. Here's an alternative definition for the serializers mapping:

```
SERIALIZERS = {
    'xml': ('application/xml', serialize_xml),
    'html': ('text/html', to_bytes(serialize_html)),
    'json': ('application/json', to_bytes(serialize_json)),
    'csv': ('text/csv', to_bytes(serialize_csv)),
}
```

This replaces decoration at the site of the function definition with decoration when building this mapping data structure. It seems potentially confusing to defer the decoration.

Serializing data into JSON or CSV formats

The JSON and CSV serializers are similar because both rely on Python's libraries to serialize. The libraries are inherently imperative, so the function bodies are strict sequences of statements.

Here's the JSON serializer:

```
import json

@to_bytes
def serialize_json(series: str, data: List[Pair]) -> str:
    """
    >>> data = [Pair(2,3), Pair(5,7)]
    >>> serialize_json( "test", data )
    b'[{"x": 2, "y": 3}, {"x": 5, "y": 7}]'
    """
    obj = [dict(x=r.x, y=r.y) for r in data]
    text = json.dumps(obj, sort_keys=True)
    return text
```

We created a list-of-dict structure and used the `json.dumps()` function to create a string representation. The JSON module requires a materialized `list` object; we can't provide a lazy generator function. The `sort_keys=True` argument value is helpful for unit testing. However, it's not required for the application and represents a bit of overhead.

Here's the CSV serializer:

```
import csv
import io

@to_bytes
def serialize_csv(series: str, data: List[Pair]) -> str:
    """
    >>> data = [Pair(2,3), Pair(5,7)]
    >>> serialize_csv("test", data)
    b'x,y\\r\\n2,3\\r\\n5,7\\r\\n'
    """
    buffer = io.StringIO()
    wtr = csv.DictWriter(buffer, Pair._fields)
    wtr.writeheader()
    wtr.writerows(r._asdict() for r in data)
    return buffer.getvalue()
```

The CSV module's readers and writers are a mixture of imperative and functional elements. We must create the writer, and properly create headings in a strict sequence. We've used the `_fields` attribute of the `Pair` namedtuple to determine the column headings for the writer.

The `writerows()` method of the writer will accept a lazy generator function. In this case, we used the `_asdict()` method of each `Pair` object to return a dictionary suitable for use with the CSV writer.

Serializing data into XML

We'll look at one approach to XML serialization using the built-in libraries. This will build a document from individual tags. A common alternative approach is to use Python introspection to examine and map Python objects and class names to XML tags and attributes.

Here's our XML serialization:

```
import xml.etree.ElementTree as XML

def serialize_xml(series: str, data: List[Pair]) -> bytes:
    """
```

```
    >>> data = [Pair(2,3), Pair(5,7)]
    >>> serialize_xml( "test", data )
    b'<series
name="test"><row><x>2</x><y>3</y></row><row><x>5</x><y>7</y></row></series>
'
    """
    doc = XML.Element("series", name=series)
    for row in data:
        row_xml = XML.SubElement(doc, "row")
        x = XML.SubElement(row_xml, "x")
        x.text = str(row.x)
        y = XML.SubElement(row_xml, "y")
        y.text = str(row.y)
    return cast(bytes, XML.tostring(doc, encoding='utf-8'))
```

We created a top-level element, `<series>`, and placed `<row>` sub-elements underneath that top element. Within each `<row>` sub-element, we've created `<x>` and `<y>` tags, and assigned text content to each tag.

The interface for building an XML document using the `ElementTree` library tends to be heavily imperative. This makes it a poor fit for an otherwise functional design. In addition to the imperative style, note that we haven't created a DTD or XSD. We have not properly assigned a namespace to our tags. We also omitted the `<?xml version="1.0"?>` processing instruction that is generally the first item in an XML document.

The `XML.tostring()` function has a type hint that states it returns `str`. This is generally true, but when we provide the `encoding` parameter, the result type changes to `bytes`. There's no easy way to formalize the idea of variant return types based on parameter values, so we use an explicit `cast()` to inform **mypy** of the actual type.

A more sophisticated serialization library could be helpful here. There are many to choose from. Visit `https://wiki.python.org/moin/PythonXml` for a list of alternatives.

Serializing data into HTML

In our final example of serialization, we'll look at the complexity of creating an HTML document. The complexity arises because in HTML, we're expected to provide an entire web page with a great deal of context information. Here's one way to tackle this HTML problem:

```
import string
data_page = string.Template("""\
```

```
<html>
<head><title>Series ${title}</title></head>
<body>
<h1>Series ${title}</h1>
<table>
<thead><tr><td>x</td><td>y</td></tr></thead>
<tbody>
${rows}
</tbody>
</table>
</body>
</html>
""")

@to_bytes
def serialize_html(series: str, data: List[Pair]) -> str:
    """
    >>> data = [Pair(2,3), Pair(5,7)]
    >>> serialize_html("test", data) #doctest: +ELLIPSIS
b'<html>...<tr><td>2</td><td>3</td></tr>\\n<tr><td>5</td><td>7</td></tr>...
    """
    text = data_page.substitute(
       title=series,
       rows="\n".join(
          "<tr><td>{0.x}</td><td>{0.y}</td></tr>".format(row)
       for row in data)
    )
    return text
```

Our serialization function has two parts. The first part is a `string.Template()` function that contains the essential HTML page. It has two placeholders where data can be inserted into the template. The `${title}` method shows where title information can be inserted, and the `${rows}` method shows where the data rows can be inserted.

The function creates individual data rows using a simple format string. These are joined into a longer string, which is then substituted into the template.

While workable for simple cases like the preceding example, this isn't ideal for more complex result sets. There are a number of more sophisticated template tools to create HTML pages. A number of these include the ability to embed the looping in the template, separate from the function that initializes serialization. Visit `https://wiki.python.org/moin/Templating` for a list of alternatives.

Tracking usage

Many publicly available APIs require the use of an *API Key*. The supplier of the API requests you to sign up and provide an email address or other contact information. In exchange for this, they provide an API Key, which activates the API.

The API Key is used to authenticate access. It may also be used to authorize specific features. Finally, it's also used to track usage. This may include throttling requests if an API Key is used too often in a given time period.

The variations in business models are numerous. For example, use of the API Key could be a billable event and charges are incurred. For other businesses, traffic must reach some threshold before payments are required.

What's important is non-repudiation of the use of the API. This, in turn, means creating API Keys that can act as a user's authentication credentials. The key must be difficult to forge and relatively easy to verify.

One easy way to create API Keys is to use a cryptographic random number to generate a difficult-to-predict key string. The `secrets` module can be used to generate unique API Key values that can be assigned to clients to track activity:

```
>>> import secrets
>>> secrets.token_urlsafe(18*size)
'kzac-xQ-BB9Wx0aQoXRCYQxr'
```

A base 64 encoding is used on the random bytes to create a sequence of characters. Using a multiple of three for the length will avoid any trailing = signs in the base 64 encoding. We've used the URL-safe base 64 encoding, which won't include the / or + characters in the resulting string. This means the key can be used as part of a URL or can be provided in a header.

The more elaborate methods won't lead to more random data. The use of `secrets` assures that no one can counterfeit a key assigned to another user.

Another choice is to use `uuid.uuid4()` to create a random **Universally Unique Identifier (UUID)**. This will be a 36-character string that has 32 hex digits and four "-" punctuation marks. A random UUID will be difficult to forge.

Another choice is to use the `itsdangerous` package to create JSON web signatures. These use a simple encryption system to make them opaque to clients, but still useful to a server. See `http://pythonhosted.org/itsdangerous/` for more information.

The RESTful web server will need a small database with the valid keys and perhaps some client contact information. If an API request includes a key that's in the database, the associated user is responsible for the request. If the API request doesn't include a known key, the request can be rejected with a simple `401 UNAUTHORIZED` response. Since the key itself is a 24-character string, the database will be rather small and can easily be cached in memory.

This small database can be a simple file that the server loads to map API Keys to authorized privileges. The file can be read at startup and the modification time checked to see if the version cached in the server is still current. When a new key is available, the file is updated and the server will re-read the file.

Ordinary log-scraping might be sufficient to show the usage for a given key. A more sophisticated application might record API requests in a separate log file or database to simplify analysis.

Summary

In this chapter, we looked at ways in which we can apply functional design to the problem of serving content with REST-based web services. We looked at the ways that the WSGI standard leads to somewhat functional overall applications. We also looked at how we can embed a more functional design into a WSGI context by extracting elements from the request for use by our application functions.

For simple services, the problem often decomposes into three distinct operations: getting the data, searching or filtering, and then serializing the results. We tackled this with three functions: `raw_data()`, `anscombe_filter()`, and `serialize()`. We wrapped these functions in a simple WSGI-compatible application to divorce the web services from the *real* processing around extracting and filtering the data.

We also looked at the way that web services' functions can focus on the *happy path* and assume that all of the inputs are valid. If inputs are invalid, the ordinary Python exception handling will raise exceptions. The WSGI wrapper function will catch the errors and return appropriate status codes and error content.

We have not looked at more complex problems associated with uploading data or accepting data from forms to update a persistent data store. These are not significantly more complex than getting data and serializing the results. However, they can be solved in a better manner.

For simple queries and data sharing, a small web service application can be helpful. We can apply functional design patterns and assure that the website code is succinct and expressive. For more complex web applications, we should consider using a framework that handles the details properly.

In the next chapter, we'll look at a few optimization techniques that are available to us. We'll expand on the `@lru_cache` decorator from `Chapter 10`, *The Functools Module*. We'll also look at some other optimization techniques that were presented in `Chapter 6`, *Recursions and Reductions*.

16
Optimizations and Improvements

In this chapter, we'll look at some optimizations that we can make to create high-performance functional programs. We will look at the following topics:

- We'll expand on using the `@lru_cache` decorator from `Chapter 10`, *The Functools Module*. We have a number of ways to implement the memoization algorithm.
- We'll also discuss how to write our own decorators. More importantly, we'll see how to use a `Callable` object to cache memoized results.
- We'll also look at some optimization techniques that were presented in `Chapter 6`, *Recursions and Reductions*. We'll review the general approach to tail recursion optimization. For some algorithms, we can combine memoization with a recursive implementation and achieve good performance. For other algorithms, memoization isn't really very helpful and we have to look elsewhere for performance improvements.
- We'll look at an extensive case study in optimizing accuracy by using the `Fraction` class.

In most cases, small changes to a program will only lead to small improvements in performance. Replacing a function with a `lambda` object will have a tiny impact on performance. If we have a program that is unacceptably slow, we often have to locate a completely new algorithm or data structure. Replacing an $\mathbf{O}(n^2)$ algorithm with one that's $\mathbf{O}(n \log n)$ is the best way to create a dramatic improvement in performance.

One place to start on redesign is `http://www.algorist.com`. This is a resource that may help to locate better algorithms for a given problem.

Memoization and caching

As we saw in `Chapter 10`, *The Functools Module*, many algorithms can benefit from memoization. We'll start with a review of some previous examples to characterize the kinds of functions that can be helped with memoization.

In Chapter 6, *Recursions and Reductions*, we looked at a few common kinds of recursions. The simplest kind of recursion is a tail recursion with arguments that can be easily matched to values in a cache. If the arguments are integers, strings, or materialized collections, then we can compare arguments quickly to determine if the cache has a previously computed result.

We can see from these examples that integer numeric calculations, such as computing factorial or locating a Fibonacci number, will be obviously improved. Locating prime factors and raising integers to powers are more examples of numeric algorithms that apply to integer values.

When we looked at the recursive version of a Fibonacci number, F_n, calculator, we saw that it contained two tail-call recursions. Here's the definition:

$$F_n = \begin{cases} 0 & \textbf{if } n = 0 \\ 1 & \textbf{if } n = 1 \\ F_{n-1} + F_{n-2} & \textbf{if } n \geq 2 \end{cases}$$

This can be turned into a loop, but any design change requires some thinking. The memoized version of a recursive definition can be quite fast and doesn't require quite so much thinking to design.

The Syracuse function is an example of the kind of function used to compute fractal values. Here's the Syracuse function, $S(n)$.

$$S(n) = \begin{cases} \frac{n}{2} & \textbf{if } n \textbf{ even} \\ 3n+1 & \textbf{if } n \textbf{ odd} \end{cases}$$

Applied recursively, there's a chain of values, C, from a given starting value, *n*.

$$C(n) = [n, S(n), S(S(n)), S(S(S(n))), \ldots] = [S^0(n), S^1(n), S^2(n), S^3(n), \ldots]$$

The Collatz conjecture is *the Syracuse function always leads to 1*. The values starting with $S(1)$ form a loop of 1, 4, 2, 1, ... Exploring the behavior of this function requires memoized intermediate results. An interesting question is locating the extremely long sequences. See `https://projecteuler.net/problem=14` for an interesting problem that requires careful use of caching.

The recursive application of the Syracuse function is an example of a function with an "attractor," where the value is attracted to 1. In some higher dimensional functions, the attractor can be a line or perhaps a fractal curve. When the attractor is a point, memoization can help; otherwise, memoization may actually be a hindrance, since each fractal value is unique.

When working with collections, the benefits of caching may vanish. If the collection happens to have the same number of integer values, strings, or tuples, then there's a chance that the collection is a duplicate and time can be saved. However, if a calculation on a collection will be needed more than once, manual optimization is best: do the calculation once and assign the results to a variable.

When working with iterables, generator functions, and other lazy objects, caching or memoization is not going to help at all. Lazy functions already do the least amount of work to provide the next value from a source sequence.

Raw data that includes measurements often use floating point values. Since an exact equality comparison between floating point values may not work out well, memoizing intermediate results may not work out well, either.

Raw data that includes counts, however, may benefit from memoization. These are integers, and we can trust exact integer comparisons to (potentially) save recalculating a previous value. Some statistical functions, when applied to counts, can benefit from using the `fractions` module instead of floating point values. When we replace `x/y` with the `Fraction(x,y)` method, we've preserved the ability to do exact value matching. We can produce the final result using the `float(some_fraction)` method.

Specializing memoization

The essential idea of memoization is so simple that it can be captured by the `@lru_cache` decorator. This decorator can be applied to any function to implement memoization. In some cases, we may be able to improve on the generic idea with something more specialized. There are a large number of potentially optimizable multivalued functions. We'll pick one here and look at another in a more complex case study.

The binomial, $\binom{n}{m}$, shows the number of ways n different things can be arranged in groups of size m. The value is as follows:

$$\binom{n}{m} = \frac{n!}{m!(n-m)!}$$

Clearly, we should cache the individual factorial calculations rather than redo all those multiplications. However, we may also benefit from caching the overall binomial calculation, too.

We'll create a `Callable` object that contains multiple internal caches. Here's a helper function that we'll need:

```
from functools import reduce
from operator import mul
from typing import Callable, Iterable

prod: Callable[[Iterable[int]], int] = lambda x: reduce(mul, x)
```

The `prod()` function computes the product of an iterable of numbers. It's defined as a reduction using the `*` operator.

Here's a `Callable` object with two caches that uses this `prod()` function:

```
class Binomial:
    def __init__(self):
        self.fact_cache = {}
        self.bin_cache = {}
    def fact(self, n: int) -> int:
        if n not in self.fact_cache:
            self.fact_cache[n] = prod(range(1, n+1))
        return self.fact_cache[n]
    def __call__(self, n: int, m: int) -> int:
        if (n, m) not in self.bin_cache:
            self.bin_cache[n, m] = (
                self.fact(n)//(self.fact(m)*self.fact(n-m)))
        return self.bin_cache[n, m]
```

We created two caches: one for factorial values and one for binomial coefficient values. The internal `fact()` method uses the `fact_cache` attribute. If the value isn't in the cache, it's computed and added to the cache. The external `__call__()` method uses the `bin_cache` attribute in a similar way: if a particular binomial has already been calculated, the answer is simply returned. If not, the internal `fact()` method is used to compute a new value.

We can use the preceding `Callable` class like this:

```
>>> binom= Binomial()
>>> binom(52, 5)
2598960
```

This shows how we can create a `Callable` object from our class and then invoke the object on a particular set of arguments. There are a number of ways that a 52-card deck can be dealt into 5-card hands. There are 2.6 million possible hands.

Tail recursion optimizations

In Chapter 6, *Recursions and Reductions*, among many others, we looked at how a simple recursion can be optimized into a `for` loop. We'll use the following simple recursive definition of factorial as an example:

$$n! = \begin{cases} 1 & \textbf{if}\, n = 0 \\ n \times (n-1)! & \textbf{if}\, n \neq 0 \end{cases}$$

The general approach to optimizing a recrusion is this:

- Design the recursion. This means the base case and the recursive cases as a simple function that can be tested and shown to be correct, but slow. Here's the simple definition:

```
def fact(n: int) -> int:
    if n == 0: return 1
    else: return n*fact(n-1)
```

- If the recursion has a simple call at the end, replace the recursive case with a `for` loop. The definition is as follows:

```
def facti(n: int) -> int:
    if n == 0: return 1
    f = 1
    for i in range(2,n):
        f = f*i
    return f
```

When the recursion appears at the end of a simple function, it's described as a tail-call optimization. Many compilers will optimize this into a loop. Python doesn't have an optimizing compiler and doesn't do this kind of tail-call transformation.

This pattern is very common. Performing the tail-call optimization improves performance and removes any upper bound on the number of recursions that can be done.

Prior to doing any optimization, it's absolutely essential that the function already works. For this, a simple `doctest` string is often sufficient. We may use annotation on our factorial functions like this:

```
def fact(n: int) -> int:
    """Recursive Factorial
    >>> fact(0)
    1
    >>> fact(1)
    1
    >>> fact(7)
    5040
    """
    if n == 0: return 1
    else: return n*fact(n-1)
```

We added two edge cases: the explicit base case and the first item beyond the base case. We also added another item that would involve multiple iterations. This allows us to tweak the code with confidence.

When we have a more complex combination of functions, we may need to execute commands such as this:

```
binom_example = """
>>> binom = Binomial()
>>> binom(52, 5)
2598960
"""

__test__ = {
    "binom_example": binom_example,
}
```

The `__test__` variable is used by the `doctest.testmod()` function. All of the values in the dictionary associated with the `__test__` variable are examined for the `doctest` strings. This is a handy way to test features that come from compositions of functions. This is also called **integration testing**, since it tests the integration of multiple software components.

Having working code with a set of tests gives us the confidence to make optimizations. We can easily confirm the correctness of the optimization. Here's a popular quote that is used to describe optimization:

> *"Making a wrong program worse is no sin."*
>
> *–Jon Bentley*

This appeared in the *Bumper Sticker Computer Science* chapter of *More Programming Pearls*, published by Addison-Wesley, Inc. What's important here is that we should only optimize code that's actually correct.

Optimizing storage

There's no general rule for optimization. We often focus on optimizing performance because we have tools such as the **Big O** measure of complexity that show us whether or not an algorithm is an effective solution to a given problem. Optimizing storage is usually tackled separately: we can look at the steps in an algorithm and estimate the size of the storage required for the various storage structures.

In many cases, the two considerations are opposed. In some cases, an algorithm that has outstandingly good performance requires a large data structure. This algorithm can't scale without dramatic increases in the amount of storage required. Our goal is to design an algorithm that is reasonably fast and also uses an acceptable amount of storage.

We may have to spend time researching algorithmic alternatives to locate a way to make the space-time trade off properly. There are some common optimization techniques. We can often follow links from Wikipedia:

`http://en.wikipedia.org/wiki/Space-time_tradeoff`.

One memory optimization technique we have in Python is to use an iterable. This has some properties of a proper materialized collection, but doesn't necessarily occupy storage. There are few operations (such as the `len()` function) that can't work on an iterable. For other operations, the memory saving feature can allow a program to work with very large collections.

Optimizing accuracy

In a few cases, we need to optimize the accuracy of a calculation. This can be challenging and may require some fairly advanced math to determine the limits on the accuracy of a given approach.

An interesting thing we can do in Python is replace floating point approximations with a `fractions.Fraction` value. For some applications, this can create more accurate answers than floating point, because more bits are used for the numerator and denominator than a floating point mantissa.

It's important to use `decimal.Decimal` values to work with currency. It's a common error to use a `float` value. When using a `float` value, additional noise bits are introduced because of the mismatch between `Decimal` values provided as input and the binary approximation used by floating point values. Using `Decimal` values prevents the introduction of tiny inaccuracies.

In many cases, we can make small changes to a Python application to switch from `float` values to `Fraction` or `Decimal` values. When working with transcendental functions, this change isn't necessarily beneficial. Transcendental functions by definition involve irrational numbers.

Reducing accuracy based on audience requirements

For some calculations, a fraction value may be more intuitively meaningful than a floating point value. This is part of presenting statistical results in a way that an audience can understand and take action on.

For example, the chi-squared test generally involves computing the χ^2 comparison between actual values and expected values. We can then subject this comparison value to a test against the χ^2 cumulative distribution function. When the expected and actual values have no particular relationship (we can call this a null relationship) the variation will be random; χ^2 the value tends to be small. When we accept the null hypothesis, then we'll look elsewhere for a relationship. When the actual values are significantly different from the expected values, we may reject the null hypothesis. By rejecting the null hypothesis, we can explore further to determine the precise nature of the relationship.

The decision is often based on the table of the χ^2 **Cumulative Distribution Function** (**CDF**) for selected χ^2 values and given degrees of freedom. While the tabulated CDF values are mostly irrational values, we don't usually use more than two or three decimal places. This is merely a decision-making tool, there's no practical difference in meaning between 0.049 and 0.05.

A widely used probability is 0.05 for rejecting the null hypothesis. This is a `Fraction` object less than 1/20. When presenting data to an audience, it sometimes helps to characterize results as fractions. A value like 0.05 is hard to visualize. Describing a relationship as having 1 chance in 20 can help people visualize the likelihood of a correlation.

Case study–making a chi-squared decision

We'll look at a common statistical decision. The decision is described in detail at `http://www.itl.nist.gov/div898/handbook/prc/section4/prc45.htm`.

This is a chi-squared decision on whether or not data is distributed randomly. To make this decision, we'll need to compute an expected distribution and compare the observed data to our expectations. A significant difference means there's something that needs further investigation. An insignificant difference means we can use the null hypothesis that there's nothing more to study: the differences are simply random variation.

We'll show how we can process the data with Python. We'll start with some backstory—some details that are not part of the case study, but often feature an **Exploratory Data Analysis** (**EDA**) application. We need to gather the raw data and produce a useful summary that we can analyze.

Within the production quality assurance operations, silicon wafer defect data is collected in a database. We may use SQL queries to extract defect details for further analysis. For example, a query could look like this:

```
SELECT SHIFT, DEFECT_CODE, SERIAL_NUMBER FROM some tables;
```

The output from this query could be a `.csv` file with individual defect details:

```
shift,defect_code,serial_number
1,None,12345
1,None,12346
1,A,12347
1,B,12348
and so on. for thousands of wafers
```

We need to summarize the preceding data. We may summarize at the SQL query level using the `COUNT` and `GROUP BY` statements. We may also summarize at the Python-application level. While a pure database summary is often described as being more efficient, this isn't always true. In some cases, a simple extract of raw data and a Python application to summarize can be faster than a SQL summary. If performance is important, both alternatives must be measured, rather than hoping that the database is fastest.

In some cases, we may be able to get summary data from the database efficiently. This summary must have three attributes: the shift, type of defect, and a count of defects observed. The summary data looks like this:

```
shift,defect_code,count
1,A,15
2,A,26
3,A,33
and so on.
```

The output will show all of the 12 combinations of shift and defect type.

In the next section, we'll focus on reading the raw data to create summaries. This is the kind of context in which Python is particularly powerful: working with raw source data.

We need to observe and compare shift and defect counts with an overall expectation. If the difference between observed counts and expected counts can be attributed to random fluctuation, we have to accept the null hypothesis that nothing interesting is going wrong. If, on the other hand, the numbers don't fit with random variation, then we have a problem that requires further investigation.

Filtering and reducing the raw data with a Counter object

We'll represent the essential defect counts as a `collections.Counter` parameter. We will build counts of defects by shift and defect type from the detailed raw data. Here's the code that reads some raw data from a `.csv` file:

```
from typing import TextIO
import csv
from collections import Counter
from types import SimpleNamespace

def defect_reduce(input_file: TextIO) -> Counter:
    rdr = csv.DictReader(input_file)
```

```
    assert set(rdr.fieldnames) == set(
        ["defect_type", "serial_number", "shift"])
    rows_ns = (SimpleNamespace(**row) for row in rdr)
    defects = (
        (row.shift, row.defect_type)
        for row in rows_ns if row.defect_type)
    tally = Counter(defects)
    return tally
```

The preceding function will create a dictionary reader based on an open file provided through the `input` parameter. We've confirmed that the column names match the three expected column names. In some cases, a file will have extra columns that must be ignored; in this case, the assertion will be something like `set(rdr.fieldnames) <= set([...])` to confirm that the actual columns are a subset of the desired columns.

We created a `types.SimpleNamespace` parameter for each row. Because the column names in the file are valid Python variable names, it's easy to transform the dictionary into a namespace object. This allows us to use slightly simpler syntax to refer to items within the row. Specifically, the next generator expression uses references such as `row.shift` and `row.defect_type` instead of the bulkier `row['shift']` or `row['defect_type']` references.

We can use a more complex generator expression to do a map–filter combination. We'll filter each row to ignore rows with no defect code. For rows with a defect code, we're mapping an expression that creates a two tuple from the `row.shift` and `row.defect_type` references.

In some applications, the filter won't be a trivial expression such as `row.defect_type`. It may be necessary to write a more sophisticated condition. In this case, it may be helpful to use the `filter()` function to apply the complex condition to the generator expression that provides the data.

Given a generator that will produce a sequence of `(shift, defect)` tuples, we can summarize them by creating a `Counter` object from the generator expression. Creating this `Counter` object will process the lazy generator expressions, which will read the source file, extract fields from the rows, filter the rows, and summarize the counts.

We'll use the `defect_reduce()` function to gather and summarize the data as follows:

```
with open("qa_data.csv") as input:
    defects = defect_reduce(input)
print(defects)
```

We can open a file, gather the defects, and display them to be sure that we've properly summarized by shift and defect type. Since the result is a `Counter` object, we can combine it with other `Counter` objects if we have other sources of data.

The `defects` value looks like this:

```
Counter({('3', 'C'): 49, ('1', 'C'): 45, ('2', 'C'): 34,
  ('3', 'A'): 33, ('2', 'B'): 31, ('2', 'A'): 26,
  ('1', 'B'): 21, ('3', 'D'): 20, ('3', 'B'): 17,
  ('1', 'A'): 15, ('1', 'D'): 13, ('2', 'D'): 5})
```

We have defect counts organized by shift (`'1', '2', or '3'`) and defect types (`'A'` through `'D'`). We'll look at alternative inputs of summarized data next. This reflects a common use case where data is available at the summary level.

Once we've read the data, the next step is to develop two probabilities so that we can properly compute expected defects for each shift and each type of defect. We don't want to divide the total defect count by 12, since that doesn't reflect the actual deviations by shift or defect type. The shifts may be more or less equally productive. The defect frequencies are certainly not going to be similar. We expect some defects to be very rare and others to be more common.

Reading summarized data

As an alternative to reading all of the raw data, we can look at processing only the summary counts. We want to create a `Counter` object similar to the previous example; this will have defect counts as a value with a key of shift and defect code. Given summaries, we simply create a `Counter` object from the input dictionary.

Here's a function that will read our summary data:

```
from typing import TextIO
from collections import Counter
import csv
def defect_counts(source: TextIO) -> Counter:
    rdr = csv.DictReader(source)
    assert set(rdr.fieldnames) == set(
        ["defect_type", "serial_number", "shift"])
    rows_ns = (SimpleNamespace(**row) for row in rdr)
    convert = map(
        lambda d: ((d.shift, d.defect_code), int(d.count)),
        rows_ns)
    return Counter(dict(convert))
```

We require an open file as the input. We'll create a `csv.DictReader()` function that helps parse the raw CSV data that we got from the database. We included an `assert` statement to confirm that the file really has the expected data.

This variation uses a `lambda` object to create a two tuple from each row. This tuple has a composite key built from shift and defect, and the integer conversion of the count. The result will be a sequence such as `((shift,defect), count), ((shift,defect), count), ...)`. When we map the `lambda` to the `row_ns` generator, we'll have a generator function that can emit the sequence of two-tuples.

We will create a dictionary from the collection of two-tuples and use this dictionary to build a `Counter` object. The `Counter` object can easily be combined with other `Counter` objects. This allows us to combine details acquired from several sources. In this case, we only have a single source.

We can assign this single source to the `defects` variable. The value looks like this:

```
Counter({('3', 'C'): 49, ('1', 'C'): 45, ('2', 'C'): 34,
 ('3', 'A'): 33, ('2', 'B'): 31, ('2', 'A'): 26,
 ('1', 'B'): 21, ('3', 'D'): 20, ('3', 'B'): 17,
 ('1', 'A'): 15, ('1', 'D'): 13, ('2', 'D'): 5})
```

This matches the detail summary shown previously. The source data, however, was already summarized. This is often the case when data is extracted from a database and SQL is used to do group-by operations.

Computing sums with a Counter object

We need to compute the probabilities of defects by shift and defects by type. To compute the expected probabilities, we need to start with some simple sums. The first is the overall sum of all defects, which can be calculated by executing the following command:

```
total = sum(defects.values())
```

This is done directly from the values in the `Counter` object assigned to the `defects` variable. This will show that there are 309 total defects in the sample set.

We need to get defects by shift as well as defects by type. This means that we'll extract two kinds of subsets from the raw defect data. The *by-shift* extract will use just one part of the `(shift,defect type)` key in the `Counter` object. The *by-type* will use the other half of the key pair.

We can summarize by creating additional `Counter` objects extracted from the initial set of the `Counter` objects assigned to the `defects` variable. Here's the by-shift summary:

```
shift_totals = sum(
    (Counter({s: defects[s, d]}) for s, d in defects),
    Counter()  # start value = empty Counter
  )
```

We've created a collection of individual `Counter` objects that have a shift, `s`, as the key and the count of defects associated with that shift `defects[s,d]`. The generator expression will create 12 such `Counter` objects to extract data for all combinations of four defect types and three shifts. We'll combine the `Counter` objects with a `sum()` function to get three summaries organized by shift.

We can't use the default initial value of 0 for the `sum()` function. We must provide an empty `Counter()` function as an initial value.

The type totals are created with an expression similar to the one used to create shift totals:

```
type_totals = sum(
    (Counter({d: defects[s, d]}) for s, d in defects),
    Counter()  # start value = empty Counter
)
```

We created a dozen `Counter` objects using the defect type, `d`, as the key instead of shift type; otherwise, the processing is identical.

The shift totals look like this:

```
Counter({'3': 119, '2': 96, '1': 94})
```

The defect type totals look like this:

```
Counter({'C': 128, 'A': 74, 'B': 69, 'D': 38})
```

We've kept the summaries as `Counter` objects, rather than creating simple `dict` objects or possibly even `list` instances. We'll generally use them as simple dicts from this point forward. However, there are some situations where we will want proper `Counter` objects instead of reductions.

Computing probabilities from Counter objects

We've read the data and computed summaries in two separate steps. In some cases, we may want to create the summaries while reading the initial data. This is an optimization that may save a little bit of processing time. We could write a more complex input reduction that emitted the grand total, the shift totals, and the defect type totals. These `Counter` objects would be built one item at a time.

We've focused on using the `Counter` instances, because they seem to allow us flexibility. Any changes to the data acquisition will still create `Counter` instances and won't change the subsequent analysis.

Here's how we can compute the probabilities of defect by shift and by defect type:

```
from fractions import Fraction
P_shift = {
    shift: Fraction(shift_totals[shift], total)
    for shift in sorted(shift_totals)
}
P_type = {
    type: Fraction(type_totals[type], total)
    for type in sorted(type_totals)
}
```

We've created two mappings: `P_shift` and `P_type`. The `P_shift` dictionary maps a shift to a `Fraction` object that shows the shift's contribution to the overall number of defects. Similarly, the `P_type` dictionary maps a defect type to a `Fraction` object that shows the type's contribution to the overall number of defects.

We've elected to use `Fraction` objects to preserve all of the precision of the input values. When working with counts like this, we may get probability values that make more intuitive sense to people reviewing the data.

The `P_shift` data looks like this:

```
{'1': Fraction(94, 309), '2': Fraction(32, 103),
 '3': Fraction(119, 309)}
```

The `P_type` data looks like this:

```
{'A': Fraction(74, 309), 'B': Fraction(23, 103),
 'C': Fraction(128, 309), 'D': Fraction(38, 309)}
```

A value such as 32/103 or 96/309 might be more meaningful to some people than 0.3106. We can easily get `float` values from `Fraction` objects, as we'll see later.

In Python 3.6, the keys in the dictionary will tend to remain in the order of the keys as found in the source data. In previous versions of Python, the order of the keys was less predictable. In this case, the order doesn't matter, but it can help debugging efforts when the keys have a predictable order.

The shifts all seem to be approximately at the same level of defect production. The defect types vary, which is typical. It appears that the defect `C` is a relatively common problem, whereas the defect `B` is much less common. Perhaps the second defect requires a more complex situation to arise.

Computing expected values and displaying a contingency table

The expected defect production is a combined probability. We'll compute the shift defect probability multiplied by the probability based on defect type. This will allow us to compute all 12 probabilities from all combinations of shift and defect type. We can weight these with the observed numbers and compute the detailed expectation for defects.

The following code calculates expected values:

```
expected = {
    (s, t): P_shift[s]*P_type[t]*total
    for t in P_type
    for s in P_shift
}
```

We'll create a dictionary that parallels the initial `defectsCounter` object. This dictionary will have a sequence of two-tuples with keys and values. The keys will be two-tuples of shift and defect type. Our dictionary is built from a generator expression that explicitly enumerates all combinations of keys from the `P_shift` and `P_type` dictionaries.

The value of the `expected` dictionary looks like this:

```
{('2', 'B'): Fraction(2208, 103),
 ('2', 'D'): Fraction(1216, 103),
 ('3', 'D'): Fraction(4522, 309),
 ('2', 'A'): Fraction(2368, 103),
 ('1', 'A'): Fraction(6956, 309),
 ('1', 'B'): Fraction(2162, 103),
```

```
('3', 'B'): Fraction(2737, 103),
('1', 'C'): Fraction(12032, 309),
('3', 'C'): Fraction(15232, 309),
('2', 'C'): Fraction(4096, 103),
('3', 'A'): Fraction(8806, 309),
('1', 'D'): Fraction(3572, 309)}
```

Each item of the mapping has a key with shift and defect type. This is associated with a `Fraction` value based on the probability of defect based on shift times, the probability of a defect based on defect type times the overall number of defects. Some of the fractions are reduced, for example, a value of 6624/309 can be simplified to 2208/103.

Large numbers are awkward as proper fractions. Displaying large values as `float` values is often easier. Small values (such as probabilities) are sometimes easier to understand as fractions.

We'll print the observed and expected times in pairs. This will help us visualize the data. We'll create something that looks like the following to help summarize what we've observed and what we expect:

```
obs exp    obs exp      obs exp     obs exp
 15 22.51    21 20.99    45 38.94    13 11.56     94
 26 22.99    31 21.44    34 39.77     5 11.81     96
 33 28.50    17 26.57    49 49.29    20 14.63    119
 74          69         128          38          309
```

This shows 12 cells. Each cell has values with the observed number of defects and an expected number of defects. Each row ends with the shift totals, and each column has a footer with the defect totals.

In some cases, we might export this data in CSV notation and build a spreadsheet. In other cases, we'll build an HTML version of the contingency table and leave the layout details to a browser. We've shown a pure text version here.

The following code contains a sequence of statements to create the contingency table shown previously:

```
print("obs exp "*len(type_totals))
for s in sorted(shift_totals):
    pairs = [
        f"{defects[s,t]:3d} {float(expected[s,t]):5.2f}"
        for t in sorted(type_totals)
    ]
    print(f"{' '.join(pairs)} {shift_totals[s]:3d}")
footers = [
    f"{type_totals[t]:3d} "
```

```
        for t in sorted(type_totals)]
    print(f"{' '.join(footers)} {total:3d}")
```

This spreads the defect types across each line. We've written enough `obs exp` column titles to cover all defect types. For each shift, we'll emit a line of observed and actual pairs, followed by a shift total. At the bottom, we'll emit a line of footers with just the defect type totals and the grand total.

A contingency table such as this one helps us to visualize the comparison between observed and expected values. We can compute a chi-squared value for these two sets of values. This will help us decide whether the data is random or whether there's something that deserves further investigation.

Computing the chi-squared value

The χ^2 value is based on $\sum_i \frac{(e_i - o_i)^2}{e_i}$, where the *e* values are the expected values and the *o* values are the observed values. In our case, we have two dimensions, shift, *s*, and defect type, *t*, which leads to $\chi^2 = \sum_s \sum_t \frac{(e_{st} - o_{st})^2}{e_{st}}$.

We can compute the specified formula's value as follows:

```
diff = lambda e, o: (e-o)**2/e

chi2 = sum(
    diff(expected[s, t], defects[s, t])
    for s in shift_totals
    for t in type_totals
)
```

We've defined a small `lambda` to help us optimize the calculation. This allows us to execute the `expected[s,t]` and `defects[s,t]` attributes just once, even though the expected value is used in two places. For this dataset, the final χ^2 value is 19.18.

There are a total of six degrees of freedom based on three shifts and four defect types. Since we're considering them to be independent, we get $(3 - 1) \times (4 - 1) = 6$. A chi-squared table shows us that anything below 12.5916 would have a 1 chance in 20 of the data being truly random. Since our value is 19.18, the data is unlikely to be random.

The cumulative distribution function for shows that a value of 19.18 has a probability of the order of 0.00387: about 4 chances in 1,000 of being random. The next step in the overall analysis is to design a follow-up study to discover the details of the various defect types and shifts. We'll need to see which independent variable has the biggest correlation with defects and continue the analysis. This work is justified because the χ^2 value indicates the effect is not simple random variation.

A side-bar question is the threshold value of 12.5916. While we can find this in a table of statistical values, we can also compute this threshold directly. This leads to a number of interesting functional programming examples.

Computing the chi-squared threshold

The essence of the χ^2 test is a threshold value based on the number of degrees of freedom and the level of uncertainty we're willing to entertain in accepting or rejecting the null hypothesis. Conventionally, we're advised to use a threshold around 0.05 (1/20) to reject the null hypothesis. We'd like there to be only 1 chance in 20 that the data is simply random and it appears meaningful. In other words, we'd like there to be 19 chances in 20 that the data reflects simple random variation.

The chi-squared values are usually provided in tabular form because the calculation involves a number of transcendental functions. In some cases, software libraries will provide an implementation of the χ^2 cumulative distribution function, allowing us to compute a value rather than look it up in a tabulation of important values.

The cumulative distribution function for an χ^2 value, *x*, and degrees of freedom, *f*, is defined as follows:

$$F(x; k) = \frac{\gamma(\frac{k}{2}, \frac{x}{2})}{\Gamma(\frac{k}{2})}$$

It's common to state the probability of being random as $p = 1 - F(\chi^2; k)$. That is, if *p > 0.05*, the data can be understood as random; the null hypothesis is true. Otherwise, the data is unlikely to be random, and further study is warranted.

The cumulative distribution is the ratio of the incomplete gamma function, $\gamma(s,z)$, and the complete gamma function, $\Gamma(x)$. The general approach to computing values for these functions can involve some fairly complex math. We'll cut some corners and implement two pretty good approximations that are narrowly focused on just this problem. Each of these functions will allow us to look at some additional functional design issues.

Both of these functions will require a factorial calculation, $n!$. We've seen several variations on the fractions theme. We'll use the following one:

```
@lru_cache(128)
def fact(k: int) -> int:
    if k < 2:
        return 1
    return reduce(operator.mul, range(2, int(k)+1))
```

This is $k! = \prod_{2 \le i \le k} i$, a product of numbers from 2 to *k* (inclusive). This implementation doesn't involve recursion. Because integers can be extremely large in Python, this has no practical upper bound on values it can compute.

Computing the incomplete gamma function

The incomplete gamma function has a series expansion. This means that we're going to compute a sequence of values and then do a sum on those values. For more information, visit `http://dlmf.nist.gov/8`.

$$\gamma(s,z) = \sum_{0 \le k < \infty} \frac{(-1)^k}{k!} \frac{z^{s+k}}{s+k}$$

This series will have an infinite sequence of terms. Eventually, the values become too small to be relevant. We can establish a small value, ε, and stop the series when the next term is less than this value.

The calculation $(-1)^k$ will yield alternating signs:

$$(-1)^0, (-1)^1, (-1)^2, (-1)^3, \ldots = 1, -1, 1, -1, \ldots$$

The sequence of terms looks like this with *s=1* and *z*=2:

```
2/1, -2/1, 4/3, -2/3, 4/15, -4/45, ..., -2/638512875
```

At some point, each additional term won't have any significant impact on the result.

When we look back at the cumulative distribution function, $F(x; k)$, we can consider working with `fractions.Fraction` values. The degrees of freedom, *k*, will be an integer divided by 2. The value, *x*, may be either a `Fraction` or a `float` value; it will rarely be a simple integer value.

When evaluating the terms of $\gamma(s, z)$, the value of $\frac{(-1)^k}{k!}$ will involve integers and can be represented as a proper `Fraction` value. The expression as a whole, however, isn't always a `Fraction` object. The value of z^{s+k} could be a `Fraction` or `float` value. The result will lead to irrational values when $s + k$ is not an integer; irrational values lead to complex-looking `Fraction` objects that approximate their value.

The use of a `Fraction` value in the gamma function may be possible, but it doesn't seem to be helpful. However, looking forward to the complete gamma function, we'll see that `Fraction` values are potentially helpful. For that reason, we'll use `Fraction` objects in this implementation even though they will have approximations of irrational values.

The following code is an implementation of the previously explained series expansion:

```
from typing import Iterator, Iterable, Callable, cast
def gamma(s: Fraction, z: Fraction) -> Fraction:

    def terms(s: Fraction, z: Fraction) -> Iterator[Fraction]:
        """Terms for computing partial gamma"""
        for k in range(100):
            t2 = Fraction(z**(s+k))/(s+k)
            term = Fraction((-1)**k, fact(k))*t2
            yield term
        warnings.warn("More than 100 terms")

    def take_until(
            function: Callable[..., bool], source: Iterable
        ) -> Iterator:
        """Take from source until function is false."""
        for v in source:
            if test(v):
                return
        yield v

    ε = 1E-8
    g = sum(take_until(lambda t: abs(t) < ε, terms(s, z)))
    # sum() from Union[Fraction, int] to Fraction
    return cast(Fraction, g)
```

We defined a `term()` function that will yield a series of terms. We used a `for` statement with an upper limit to generate only 100 terms. We could have used the `itertools.count()` function to generate an infinite sequence of terms. It seems slightly simpler to use a loop with an upper bound.

We computed a potentially irrational z^{s+k} value and created a `Fraction` value from this value by itself. The division of $\frac{z^{s+k}}{s+k}$ will involve division of two `Fractions`, creating a new `Fraction` value for the `t2` variable. The value for the `term` variable will then be a product of two `Fraction` objects.

We defined a `take_until()` function that takes values from an iterable, until a given function is true. Once the function becomes true, no more values are consumed from the iterable. We also defined a small threshold value, ε, of 10^{-8}. We'll take values from the `term()` function until the values are less than ε. The sum of these values is an approximation to the partial `gamma` function.

Note that the threshold variable is the Greek letter epsilon. Python 3 allows variable names to use any Unicode letter.

Here are some test cases we can use to confirm that we're computing this properly:

- $\gamma(1, 2) = 1 - e^{-2} \approx 0.8646647$
- $\gamma(1, 3) = 1 - e^{-3} \approx 0.9502129$
- $\gamma(\frac{1}{2}, 2) = \sqrt{\pi} \times \text{erf}(\sqrt{2}) \approx 1.6918067$

The error function, `erf()`, is available in the Python math library. We don't need to develop an approximation for this.

Our interest is narrowly focused on the chi-squared distribution. We're not generally interested in the incomplete `gamma` function for other mathematical purposes. Because of this, we can narrow our test cases to the kinds of values we expect to be using. We can also limit the accuracy of the results. Most chi-squared tests involve three digits of precision. We've shown seven digits in the test data, which is more than we might properly need.

Computing the complete gamma function

The complete gamma function is a bit more difficult to implement than the incomplete gamma function. There are a number of different approximations. For more information, visit `http://dlmf.nist.gov/5`. There's a version available in the Python `math` library. It represents a broadly useful approximation that is designed for many situations.

We're not interested in a completely general implementation of the complete gamma function. We're interested in just two special cases: integer values and halves. For these two special cases, we can get exact answers, and don't need to rely on an approximation.

For integer values, $\Gamma n = (n-1)!$. The complete gamma function for integers can rely on the factorial function we defined previously.

For values which are halves, there's a special form: $\Gamma(\frac{1}{2}+n) = \frac{(2n)!}{4^n n!}\sqrt{\pi}$. This includes an irrational value, $\sqrt{\pi}$, so we can only represent this approximately using `float` or `Fraction` objects.

If we use proper `Fraction` values, then we can design a function with a few simple cases: an `integer` value, a `Fraction` value with `1` in the denominator, and a `Fraction` value with 2 in the denominator. We can use the `Fraction` value as shown in the following code:

```
sqrt_pi = Fraction(677_622_787, 382_307_718)

from typing import Union
def Gamma_Half(
        k: Union[int, Fraction]
    ) -> Union[int, Fraction]:
    if isinstance(k, int):
        return fact(k-1)
    elif isinstance(k, Fraction):
        if k.denominator == 1:
            return fact(k-1)
        elif k.denominator == 2:
            n = k-Fraction(1, 2)
            return fact(2*n)/(Fraction(4**n)*fact(n))*sqrt_pi
    raise ValueError(f"Can't compute Γ({k})")
```

We called the function `Gamma_Half` to emphasize that this is only appropriate for whole numbers and halves. For integer values, we'll use the `fact()` function that was defined previously. For `Fraction` objects with a denominator of 1, we'll use the same `fact()` definition.

For the cases where the denominator is 2, we can use the more complex *closed form* value. We used an explicit `Fraction()` function for the value $4^n n!$. We've also provided a `Fraction` approximation for the irrational value $\sqrt{\pi}$.

Here are some test cases:

- $\Gamma(2) = 1$
- $\Gamma(5) = 24$
- $\Gamma(\frac{1}{2}) = \sqrt{\pi} \approx 1.7724539 \approx \frac{582,540}{328,663}$
- $\Gamma(\frac{3}{2}) = \frac{\sqrt{\pi}}{2} \approx 0.8862269 \approx \frac{291,270}{328,663}$

These can also be shown as proper `Fraction` values. The presence of irrational numbers (square roots, and π) tends to create large, hard-to-read fractions. We can use something like this to create easier-to-read fractions:

```
>>> g = Gamma_Half(Fraction(3, 2))
>>> g.limit_denominator(2_000_000)
Fraction(291270, 328663)
```

This provides a value where the denominator has been limited to be less than two million; this provides pleasant-looking, six-digit numbers that we can use for unit test purposes.

Computing the odds of a distribution being random

Now that we have the incomplete `gamma` function, `gamma`, and the complete `gamma` function, `Gamma_Half`, we can compute the χ^2 `CDF` values. This value shows us the odds of a given value being random or having some possible correlation.

The function itself is quite small:

```
def cdf(x: Union[Fraction, float], k: int) -> Fraction:
    """X² cumulative distribution function.
    :param x: X² value, sum (obs[i]-exp[i])**2/exp[i]
        for parallel sequences of observed and expected values.
    :param k: degrees of freedom >= 1; often len(data)-1
    """
    return (
        1 -
```

```
        gamma(Fraction(k, 2), Fraction(x/2)) /
        Gamma_Half(Fraction(k, 2))
    )
```

This function includes some `docstring` comments to clarify the parameters. We created proper `Fraction` objects from the degrees of freedom and the chi-squared value, x. This will be a `float` value for the `x` parameter to a `Fraction` object, allowing some flexibility to match examples that are entirely done in float-point approximations.

We can use `Fraction(x/2).limit_denominator(1000)` to limit the size of the `x/2Fraction` method to a respectably small number of digits. This will compute a correct `CDF` value, but won't lead to gargantuan fractions with dozens of digits.

Here is some sample data called from a table of x^2. Visit `http://en.wikipedia.org/wiki/Chi-squared_distribution` for more information.

To compute the correct `CDF` values, execute the following kinds of commands:

```
>>> round(float(cdf(0.004, 1)), 2)
0.95
>>> cdf(0.004, 1).limit_denominator(100)
Fraction(94, 99)
>>> round(float(cdf(10.83, 1)), 3)
0.001
>>> cdf(10.83, 1).limit_denominator(1000)
Fraction(1, 1000)
>>> round(float(cdf(3.94, 10)), 2)
0.95
>>> cdf(3.94, 10).limit_denominator(100)
Fraction(19, 20)
>>> round(float(cdf(29.59, 10)), 3)
0.001
>>> cdf(29.59, 10).limit_denominator(10000)
Fraction(8, 8005)
```

Given χ^2 and a number of degrees of freedom, our `CDF` function produces the same results as a widely used table of values. The first example shows how likely an χ^2 of 0.004 with one degree of freedom would be. The second example shows how likely an χ^2 of 10.38 would be with one degree of freedom. A small χ^2 value means there's little difference between expected and observed outcomes.

Here's an entire row from an χ^2 table, computed with a simple generator expression:

```
>>> chi2 = [0.004, 0.02, 0.06, 0.15, 0.46, 1.07, 1.64, ...     2.71, 3.84,
6.64, 10.83]
>>> act = [round(float(x), 3)
...     for x in map(cdf, chi2, [1]*len(chi2))]
>>> act
[0.95, 0.888, 0.806, 0.699, 0.498, 0.301, 0.2, 0.1, 0.05, 0.01, 0.001]
```

These values show the relative likelihood of the given χ^2 for outcomes with one degree of freedom. Our computation has some tiny discrepancies in the third decimal place when compared with the published results. This means we can use our CDF computation instead of looking up χ^2 values in a standard statistical reference.

The `CDF()` function gives us the probability of a χ^2 value being due to random chance. From a published table, the 0.05 probability for six degrees of freedom has a χ^2 value 12.5916. Here's the output from this `CDF()` function, showing a good agreement with published results:

```
>>> round(float(cdf(12.5916, 6)), 2)
0.05
```

Looking back at our previous example, the actual value we got for χ^2 in the example was 19.18. Here's the probability that this value is random:

```
>>> round(float(cdf(19.18, 6)), 5)
0.00387
```

This probability is 3/775, with the denominator limited to 1,000. Those are not good odds of the data being random. This means that the null hypothesis can be rejected, and more investigation can be done to determine the likely cause of the differences.

Functional programming design patterns

There are a number of common design patterns for functional programming. These are typical approaches to functional programming that are used in a variety of contexts.

Note the important distinction from object-oriented design patterns. Many OO design patterns are designed to make management of state more explicit, or aid in composition of complex, emergent behavior. For functional design patterns, the focus is almost always on creating complex behavior from simpler pieces.

There are many common functional design approaches shown throughout this book. Most have not been presented with a particular name or story. In this section, we'll review a number of these patterns.

- **Currying**: This can be called a partial function application and is implemented with the `partial()` function in the `functools` module. The idea is to create a new function based on an existing function plus some (but not all) of the function's arguments.
- **Closures**: In Python, it's very easy to define a function that returns another function. When the returned function includes variables bound by the outer function, this is a closure. This is often done when a function returns a lambda object or a generator expression. It's also done as part of creating a parameterized decorator.
- **Pure functions**: These are the common stateless functions. In Python, we may also work with impure functions to deal with stateful input and output. Additionally, system services and random number generators are examples of impure functions. A good functional design will tend to emphasize pure functions to the furthest extent possible, avoiding the `global` statement.
- **Function composition**: The `itertools` library contains a number of tools for functional composition. In previous chapters, we also looked at ways to use decorators for functional composition. In many cases, we'll want to create callable objects so that we can bind functions together at runtime.
- **Higher-order functions**: Python has a number of built-in functions that use other functions, these include `map()`, `filter()`, `min()`, `max()`, and `sorted()`. Additionally, libraries such as `functools` and `itertools` contain other examples.
- **Map-reduce algorithms**: They are easily built using the higher-order functions. In Python, these amount to a variation on `reduce(f, map(g, data))`. We can use a function, `f()`, to handle reduction, and a function, `g()`, to perform an item-by-item mapping. Common examples of reductions include `sum()`, as well as many of the functions in the `statistics` library.

- **Lazy ("non-strict") evaluation:** This is exemplified by Python generator expressions. An expression such as `(f(a) for a in S)` is lazy and will only evaluate `f(a)` as values are consumed by some client operation. In many examples, we've used the `list()` function to consume values from a lazy generator.
- **Monads:** Imposing ordering is generally needless in Python because ordering of operations is unavoidable. We can use the `pymonad` library to provide some explicit syntax that can clearly show how ordering should be implemented within more complex expressions. This is helpful for input and output, but also for complex simulations where behavior is stateful.

In addition to these common functional programming design patterns, there are some additional techniques that are part of doing functional programming in Python:

- **Transforming tail recursion into a `for` statement**: Python imposes an upper limit on recursion and there are rare cases where loops will allow us to exceed this limit. More importantly, recursion involves overhead in managing stack frames, which a `for` statement avoids.
- **Iterable functions:** Using the `yield from` statement makes it very easy to create functions that are iterable collections of results from other functions. Using iterable results facilitates functional composition.
- Python decorators and callable objects can behave as *functors*. In the ML-like languages, functors are used to take type definitions as parameters. In Python, type definitions are often class-based, and it's sensible to combine these with callable objects or decorators.

All of these functional design patterns can be characterized as typical or common ways to design and implement functional programming. Any kind of design that's frequently repeated forms a pattern we can learn from and use for our own software design.

Summary

In this chapter, we looked at three optimization techniques. The first technique involves finding the right algorithm and data structure. This has more impact on performance than any other single design or programming decision. Using the right algorithm can easily reduce runtimes from minutes to fractions of a second. Changing a poorly used sequence to a properly used mapping, for example, may reduce runtime by a factor of 200.

We should generally optimize all of our recursions to be loops. This will be faster in Python and it won't be stopped by the call stack limit that Python imposes. There are many examples of how recursions are flattened into loops in other chapters, primarily, Chapter 6, *Recursions and Reductions*. Additionally, we may be able to improve performance in two other ways. First, we can apply memoization to cache results. For numeric calculations, this can have a large impact; for collections, the impact may be less. Secondly, replacing large materialized data objects with iterables may also improve performance by reducing the amount of memory management required.

In the case study presented in this chapter, we looked at the advantage of using Python for exploratory data analysis—the initial data acquisition including a little bit of parsing and filtering. In some cases, a significant amount of effort is required to normalize data from various sources. This is a task at which Python excels.

The calculation of an χ^2 value involved three `sum()` functions: two intermediate generator expressions, and a final generator expression to create a dictionary with expected values. A final `sum()` function created the statistic. In under a dozen expressions, we created a sophisticated analysis of data that will help us accept or reject the null hypothesis.

We also evaluated some complex statistical functions: the incomplete and the complete gamma functions, $\gamma(s, z)$ and $\Gamma(k)$. The incomplete gamma function involves a potentially infinite series; we truncated this and summed the values. The complete gamma function has some potential complexities, but those don't happen to apply to this situation.

Using a functional approach, we can write succinct and expressive programs that accomplish a great deal of processing. Python isn't a properly functional programming language. For example, we're required to use some imperative programming techniques. This limitation forces us away from purely functional recursions. We gain some performance advantage, since we're forced to optimize tail recursions into explicit loops.

We also saw numerous advantages of adopting Python's hybrid style of functional programming. In particular, the use of Python's higher-order functions and generator expressions give us a number of ways to write high-performance programs that are often quite clear and simple.

Other Books You May Enjoy

If you enjoyed this book, you may be interested in these other books by Packt:

Learning Concurrency in Python
Elliot Forbes

ISBN: 978-1-78728-537-8

- Explore the concept of threading and multiprocessing in Python
- Understand concurrency with threads
- Manage exceptions in child threads
- Handle the hardest part in a concurrent system — shared resources
- Build concurrent systems with Communicating Sequential Processes (CSP)
- Maintain all concurrent systems and master them
- Apply reactive programming to build concurrent systems
- Use GPU to solve specific problems

Python Microservices Development
Tarek Ziadé

ISBN: 978-1-78588-111-4

- Explore what microservices are and how to design them
- Use Python 3, Flask, Tox, and other tools to build your services using best practices
- Learn how to use a TDD approach
- Discover how to document your microservices
- Configure and package your code in the best way
- Interact with other services
- Secure, monitor, and scale your services
- Deploy your services in Docker containers, CoreOS, and Amazon Web Services

Leave a review - let other readers know what you think

Please share your thoughts on this book with others by leaving a review on the site that you bought it from. If you purchased the book from Amazon, please leave us an honest review on this book's Amazon page. This is vital so that other potential readers can see and use your unbiased opinion to make purchasing decisions, we can understand what our customers think about our products, and our authors can see your feedback on the title that they have worked with Packt to create. It will only take a few minutes of your time, but is valuable to other potential customers, our authors, and Packt. Thank you!

Index

D

E

F

G

H

I

J

K

L

M

N

O

P

U

W

X

Y

Z